T0225061

SpringerBriefs in Electrical and Computer Engineering

Series Editors

Woon-Seng Gan, School of Electrical and Electronic Engineering, Nanyang Technological University, Singapore, Singapore

C.-C. Jay Kuo, University of Southern California, Los Angeles, CA, USA

Thomas Fang Zheng, Research Institute of Information Technology, Tsinghua University, Beijing, China

Mauro Barni, Department of Information Engineering and Mathematics, University of Siena, Siena, Italy

SpringerBriefs present concise summaries of cutting-edge research and practical applications across a wide spectrum of fields. Featuring compact volumes of 50 to 125 pages, the series covers a range of content from professional to academic. Typical topics might include: timely report of state-of-the art analytical techniques, a bridge between new research results, as published in journal articles, and a contextual literature review, a snapshot of a hot or emerging topic, an in-depth case study or clinical example and a presentation of core concepts that students must understand in order to make independent contributions.

More information about this series at http://www.springer.com/series/10059

Ilya Gertsbakh · Yoseph Shpungin

Network Reliability

A Lecture Course

 Springer

Ilya Gertsbakh
Department of Mathematics
Ben Gurion University
Beer-Sheva, Israel

Yoseph Shpungin
Software Engineering Department
Shamoon College of Engineering
Beer-Sheva, Israel

ISSN 2191-8112   ISSN 2191-8120   (electronic)
SpringerBriefs in Electrical and Computer Engineering
ISBN 978-981-15-1457-9   ISBN 978-981-15-1458-6   (eBook)
https://doi.org/10.1007/978-981-15-1458-6

This Springer imprint is published by the registered company Springer Nature Singapore Pte Ltd.
The registered company address is: 152 Beach Road, #21-01/04 Gateway East, Singapore 189721, Singapore

Preface

The reader interested in learning about network reliability may read these lectures in two ways—as a short or as a full course.

In the first case, he/she starts with Chap. 2 which gives a formal definition of a network and describes various ways of defining network reliability. Then goes Chap. 3 which describes elementary methods of calculating network reliability. Next Chap. 4 describes a universal and rather simple working tool—so-called crude Monte Carlo. This method is not the best one but can work in any situation. Chapter 4 presents also a typical pseudocode for a Monte Carlo computer program. The short course is concluded by Chap. 5, which describes applications of theoretical material about component importance. Chapters 2–5 already enable to solve, perhaps in a nonoptimal way, a wide range of network reliability problems. The reader can also look through Chap. 8 if she/he is interested in some new applications of network reliability theory to practical problems.

For a full course, the reader starts with Chap. 1 and goes through all other chapters.

The contents of these lectures correspond to Network Reliability course delivered in Sami Shamoon Engineering College (Beer-Sheva, Israel) for M.Sc. students specializing in software engineering. It was delivered in 2013–2018 years as 1-semester course, 4 hours weekly. The prerequisites are elementary probability theory and computer programming.

We strongly recommend to the lecturers of this course to include in the final mark also the results of two–three homeworks devoted to network simulation and reliability analysis.

We tried to write these lectures in a simple and reader-friendly way. We avoided complex proofs and tried, wherever possible, to rely on intuition and explain formal statements by numerical examples.

George Box used to cite the aphorism: "all models are wrong; some models are useful." We hope that some models presented in this book might be useful to reliability researchers involved in network study and design and to reliability engineers interested in applications of the theory to practical calculations of network reliability parameters.

Beer-Sheva, Israel   Ilya Gertsbakh
September 2019   Yoseph Shpungin

Contents

Notations and Abbreviations

up, down	Element states in the system (in the network)
network element	Edge or node of the network subject to failure
UP, DOWN	States of the network
c.d.f., CDF	Cumulative distribution function
τ, X, Y, Z	Random variables (r.v.s.)
CD-spectrum	Cumulative destruction spectrum
D-spectrum	Destruction spectrum
$X \sim \text{Exp}(\lambda)$	r.v. X is exponentially distributed with parameter λ
r.e., rel.err.	Relative error
$\mathcal{N} = (V, E, T)$	Network with node (vertex) set V, edge (link) set E and terminal set T
p, q	Probability that an element is in state *up* and *down*, respectively
$\mathbf{x} = (x_1, x_2, \ldots, x_n)$	System (network) state vector; $x_i = 1$, or $x_i = 0$ if element i is *up* or *down*, respectively
$C(x)$	Number of system failure sets with x elements *down*
$Y \sim \mathbf{B}(n, p)$	r.v. Y has binomial distribution
$R(p_1, \ldots, p_n)$	Network reliability as a function of elements reliability
BIM	Birnbaum importance measure
BIM_j	Birnbaum importance measure of element j

Chapter 1
Probability-Reminder

Abstract This chapter contains material which you certainly have learned some time ago and which is familiar to you. No doubt, you have met some elements of algebra of events, random variables, distribution function and have heard about binomial distribution. Among the material of this chapter is a fundamental Law of Total Probability which will be used further in our exposition. Maybe, the section of order statistics is new for you. This is not complex but very important material which will be used in introducing a cumulative destruction spectrum. Our advice–ignore the algebra and concentrate on the final results. The last section contains a few topics from statistics. They will be used to evaluate the relative error of Monte Carlo estimation procedures.

Keywords Random variables · Binomial, exponential distribution · Order statistics · Law of total probability

1.1 Algebra of Events

Probability theory usually starts with the Algebra of Events. We will present a reminder of this material illustrated by an example of three cities Arno, Barno and Carno located in an area subject to often tornadoes. These cities are connected by roads and the tornadoes usually damage bridges on these roads (see the map below). These bridges are named A, B and C. Local observers came to the conclusion that the probabilities that these bridges will **not** be damaged during tornado period are $P(A) = 0.5$, $P(B) = 0.7$ and $P(C) = 0.4$. A long period of observations led the observers to conclude that bridge damages are *independent* events. Then, for example, $P(A \cap B) = 0.35$, $P(A \cap B \cap C) = 0.14$.   Let us compute the probability that the connection between Arno and Carno will *not* be damaged after the tornado period (Fig. 1.1).

  Connection between Arno and Carno, as we see from the map, will be open in one of three cases:

all three bridges A, B, C are not damaged; this is the event $D = A \cap B \cap C$.
A is damaged, B and C are not damaged; this is the event $E = \overline{A} \cap B \cap C$.

© The Author(s), under exclusive license to Springer Nature Singapore Pte Ltd. 2020
I. Gertsbakh and Y. Shpungin, *Network Reliability*,
SpringerBriefs in Electrical and Computer Engineering,
https://doi.org/10.1007/978-981-15-1458-6_1

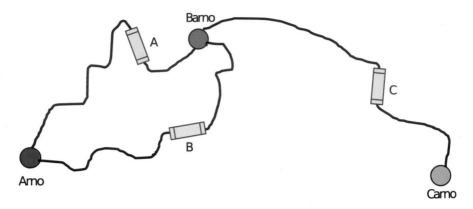

Fig. 1.1 Bridges of the tornado area

B is damaged, A and C are not damaged; this is the event $F = \overline{B} \cap A \cap C$.

Since A, B, C are independent events, the probabilities of D, E, F are:
$P(D) = P(A \cap B \cap C) = 0.5 \cdot 0.7 \cdot 0.4 = 0.14$, $P(E) = P(\overline{A} \cap B \cap C) = 0.5 \cdot 0.3 \cdot 0.4 = 0.14$, and $P(F) = (P(A \cap \overline{B} \cap C) = 0.5 \cdot 0.3 \cdot 0.4 = 0.06$. The events D, E, F are mutually exclusive and therefore the probability that there will be a connection from Arno to Carno equals $P(D \cup E \cup F) = P(D) + P(E) + P(F) = 0.14 + 0.14 + 0.06 = 0.34$.

Let us compute the conditional probability of the event $A \cap B \cap C$ *given* the event $A \cap C$. By well-known formula of the conditional probability

$$P(A \cap B \cap C | A \cap C) = \frac{P(A \cap B \cap C)}{P(A \cap C)} = 0.5 \cdot 0.7 \cdot 0.4/(0.5 \cdot 0.4) = 0.7.$$

1.2   Random Variables

One of the central notions in probability theory is *random variable*. We will distinguish *discrete* and *continuous* random variables.

Discrete random variable. Imagine a random experiment whose outcome can be one of the numbers x_i from the following finite or infinite collection

$$x_1, x_2, x_3, \ldots, x_n, \ldots$$

The outcome x_i appears with probability p_i. This fact is formally written as

$$P(X = x_i) = p_i. \tag{1.2.1}$$

Here capital X is the name of the random variable. Of course,

$$\sum_{i=1}^{\infty} p_i = 1, \; p_i \geq 0. \tag{1.2.2}$$

Example. Binomial random variable.

The simplest and at the same time the most useful random variable is so-called Bernoulli (or binary) random variable (r.v.) Y for which the list of the outcomes has only two values–0 and 1 and the corresponding probabilities are

$$P(Y = 0) = q, \; P(Y = 1) = 1 - q = p.$$

In probability slang getting zero is called "failure" and getting 1 is called "success". Next important notion is *independent random variables*.

Suppose we have a family of binary r.v.s

$$Y_1, Y_2, \ldots, Y_m$$

Denote by x_i the possible value that Y_i can take (i.e. 0 or 1). Suppose that the **joint probability**

$$P(Y_1 = x_1 \cap Y_2 = x_2 \cap \cdots \cap Y_m = x_m) =$$

$$P(Y_1 = x_1) \cdot P(Y_2 = x_2) \cdot \cdots \cdot P(Y_m = x_m)$$

This last formula looks a bit frightening. In fact, it says that the joint probability is just a product of particular probabilities. For example, we have 5 *independent* Bernoulli random experiments $Y_1, \ldots, Y_5$ with the same probabilities p for success and $q = 1 - p$ for failure. Suppose that the first three ended with failure and the last two with success. Then by the above formula

$$P(Y_1 = Y_2 = Y_3 = 0 \cap Y_4 = Y_5 = 1) = q^3 p^2.$$

Now we are ready to introduce an important **binomial** random variable. Suppose we have n independent Bernoulli experiments $Y_1, Y_2, \ldots, Y_n$ having the same probability p of success. Denote by

$$Y = Y_1 + Y_2 + \cdots + Y_n.$$

Y is the total number of successes in n independent Bernoulli experiments.

We present here without proof the following formula

$$P(Y = k) = \frac{n!}{k!(n-k)!} p^k q^{n-k} \tag{1.2.3}$$

Fig. 1.2 Uniform
distribution density function
$f(t)$ and the corresponding
cumulative function $F(t)$

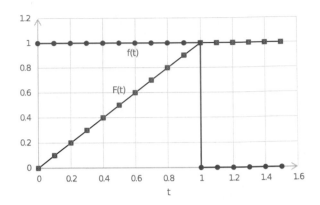

We will use the following shorthand notation for the binomial random variable

$$Y \sim \mathbf{B}(n, p).$$

For binomial random variable, the probability or the "probability mass" is concentrated in finite number of points. This is why we call it *discrete* random variable.

Let us now consider a random experiment the outcome of which Z can be *any number* in a finite or infinite interval. The location of the probability for that case is determined by a non negative function $f(x)$ having the following property.

$$P(A < Z \le B) = \int_A^B f(x)dx. \tag{1.2.4}$$

$f(x)$ is called *probability density* function. Obviously, $\int_{-\infty}^{\infty} f(x)dx = 1$.

Consider, for example, so-called uniform random variable which has density function $f(x) = 1$ for x in the interval $[0, 1]$ and zero otherwise. This random variable will be denoted as $Z \sim U(0, 1)$.

Figure 1.2(left) shows the density function of this random variable. Figure 1.2 (right) shows so-called *cumulative* or *distribution* function (cdf) of this random variable. The cumulative distribution function $F(t)$ of a random variable having density function $f(x)$ is determined by the following formula:

$$F(t) = \int_{-\infty}^{t} f(x)dx \tag{1.2.5}$$

A very important random variable for our lecture course will be so-called exponential random variable. Its density is

$$f(t) = \lambda e^{-\lambda t}, t \ge 0, \tag{1.2.6}$$

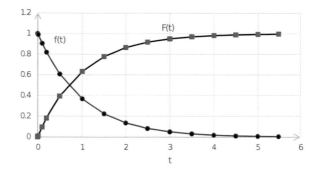

Fig. 1.3 Exponential density function $f(t)$ and cumulative function $F(t)$. For $t < 0$, $f(t) = F(t) = 0$

where λ is a positive parameter. The corresponding cumulative distribution function is

$$F(t) = 0 \text{ for } t < 0 \text{ and } F(t) = 1 - e^{-\lambda t} \text{ for } t \geq 0. \tag{1.2.7}$$

Figure 1.3 shows the density and the cumulative distribution function for this random variable.

We will use the following shorthand notation for exponentially distributed random variable:

$$V \sim Exp(\lambda)$$

The average (the mean value) of exponential random variable is

$$E[V] = \int_0^\infty x\lambda e^{-\lambda x} dx = 1/\lambda. \tag{1.2.8}$$

Exponential random variable has many amazing properties about which we will tell you later.

The following property of the exponential random variable is called *memoryless property*. Let us compute the *conditional* probability that $V \leq (t + x)$ *given* that it exceeds t. By the conditional probability formula

$$P(A|B) = P(A \cap B)/P(B)$$

we obtain

$$P(V \leq t + x | V > t) = \frac{P(t < V \leq t + x)}{P(V > t)}$$

$$= \frac{(1 - e^{-\lambda(t+x)}) - (1 - e^{-\lambda t})}{e^{-\lambda t}} = 1 - e^{-\lambda x}. \tag{1.2.9}$$

We see from here that the fact that $V > t$ does not influence the exponential probability density function.

For a discrete random variable having values $x_1, x_2, \ldots x_n$ with probabilities $p_1, p_2, \ldots p_n$, $p_i = P(X = x_i)$, the mean value is defined by the formula

$$Mean = E[X] = \sum_{i=1}^{n} x_i p_i \qquad (1.2.10)$$

If probabilities p_i can be considered as mass located at x_i then $E[X]$ is the center of gravity of these mass.

So far we know only one discrete random variable $X \sim \mathbf{B(n,p)}$. For it,

$$E[X] = np$$

1.3 The Law of Total Probability

Suppose that the sample space Ω is a union of non intersecting and non empty events $B_1, B_2, \ldots, B_k$: $\Omega = \cup_{i=1}^{k} B_i$. Let A be any event, Then it may be represented in the following form:

$$A = A \cap B_1 \cup A \cap B_2 \cup \cdots \cup A \cap B_k.$$

Representing each probability of $A \cap B_j$ as

$$P(A \cap B_j) = P(B_j)P(A|B_j)$$

we arrive at the formula representing the Law of Total Probability:

$$P(A) = \sum_{i=1}^{k} P(B_i) \cdot P(A|B_i). \qquad (1.3.1)$$

1.4 Order Statistics

Let $X_1, X_2, \ldots X_n$ be independent, positive,identically distributed continuous random variables. In our exposition, we typically use such random variables to describe network element lifetimes. The cumulative distribution function of X_i will be denoted by $G(t)$, i.e. $P(X_i \leq t) = G(t)$.

Imagine that we have a generator of each of these random variables and observe one random exemplar of each variable. Denote by $X_{(1)}$ the smallest of these values, by $X_{(2)}$ the second smallest value, and so on. $X_{(n)}$ will be the largest of our observations:

$$X_{(1)} \le X_{(2)} \le \cdots X_{(n)}. \tag{1.4.1}$$

Definition 1.1 The rth smallest value in (1.4.1) is called the rth order statistic, $r = 1, 2, \ldots n.\#$

In further exposition, we will need the c.d.f. of the rth order statistics, i.e. $P(X_{(r)} \le t) = G_{(r)}(t)$. $X_{(r)}$ is less or equal t if at least r of n random variables X_i are $\le t$. This immediately leads us to the formula:

$$P(X_{(r)} \le t) = G_{(r)}(t) = \sum_{j=r}^{n} \frac{n!}{j!(n-j)!} [G(t)]^j [1 - G(t)]^{(n-j)}. \tag{1.4.2}$$

Imagine now the following experiment. We have a network (a formal definition of a network and its failure can be found in Sect. 2.1), and we number its elements subject to failure by $i = 1, 2, \ldots, n$. Assign to an element with number r, a random variable X_r. $X_1, X_2, \ldots, X_n$ are i.i.d. random variables. Observe the **state of the network** after each failure of one of its elements. With some probability f_1 the network will fail when the first element fails ("first" means first in time, with the shortest lifetime), with probability f_2 network fails when the element having the second shortest lifetime fails, and so on. By f_k the network failure coincides with the kth order statistics of element lifetime.

Denote by τ network lifetime and by $G(t)$ the c.d.f. of τ: $P(\tau \le t) = G(t)$. Obviously, network lifetime coincides with one of the random moments of the failure of its elements. Using the Law of Total Probability, Samaniego [1] derived the following very useful formula:

$$P(\tau \le t) = G(t) = \sum_{j=1}^{n} f_j G_{(j)}(t). \tag{1.4.3}$$

Therefore, network lifetime is equal to a linear combination of element lifetime order statistics. Following Samaniego [1], the coefficients $(f_1, f_2, \ldots, f_n)$ are called *signature*. A few years before, Lomonosov in [2] suggested the term "internal distribution" (ID). The paper [2] presents the ID's for the family of complete graphs $K_5, \ldots, K_{11}$ with 5–11 nodes, with the definition that network failure is defined as the loss of overall connectivity.

Let us now substitute the expression for $G_{(r)}$ from (1.4.2) into (1.4.3). After some algebra, by changing the order of summation, we obtain:

$$P(\tau \le t) = G(t) = \sum_{k=1}^{k=n} \sum_{j=1}^{k} f_j \frac{n!}{k!(n-k)!} [G(t)]^k [1 - G(t)]^{n-k}. \tag{1.4.4}$$

1.5   Average, Variance

Right now we will be entering a part of statistical inference needed for our studies. Suppose we don't know p and we want to **estimate** it using repeated independent observations on the same binomial random variable. Here we come to so-called *average* of several independent observations. Suppose we repeat the experiment which produces a replica of r.v. $X \sim \mathbf{B}(n, p)$. Let us produce M replicas of X. Denote them $X_1, X_2, \ldots X_M$ and take their average

$$\hat{\mu} = \bar{X} = \frac{X_1 + X_2 + \cdots + X_M}{M}$$

All X_i have the same mean value $E(X_i) = np$, and also $E[\bar{X}] = np$. Thus $\hat{\mu}$ is so called **unbiased estimator** of $\mu = E[X] = np$.

Here we already are in more complicated situation. We have a sum of several random variables divided by their number M. Expressions like this are studied by mathematical statistics. It turns out that a very important fact takes place: as M grows, $\hat{\mu} \to np$ in the sense that the values of the variable $\hat{\mu}$, with high probability, are close to np.

Suppose $X_i \sim B(1, p)$. Then as $M \to \infty$

$$\frac{X_1 + X_2 + \cdots X_M}{M} = \hat{\mu} \to p \qquad (1.5.1)$$

In other words: if we have a binary random variable with probability of success p in a single experiment, then the average number of M random replicas of these random variables is approaching p if the number of experiments M increases.

We would like to know how close is this variable $\hat{\mu}$ to the unknown value p. The answer is supplied by the so called central limit theorem. Let us present here without proof the relevant result which we will use in the analysis of our Monte Carlo experiments.

Theorem 1.1 *With probability 0.95, the average value of M independent binomial experiments $\hat{\mu}$ lies within the interval $[LB, UB]$, where*

$$LB = \hat{\mu} - 1.96\sqrt{\hat{\mu}(1 - \hat{\mu})/M}, \, UB = \hat{\mu} + 1.96\sqrt{\hat{\mu}(1 - \hat{\mu})/M}, \#$$

Theorem 1.2 *Confidence interval for mean value of random variable.*
Suppose we observe a sample $(x_1, x_2, \ldots, x_N)$ of N values of a random variable X. X has mean value μ which is unknown for us. Then the 95 percent confidence interval for μ is

$$LB = \bar{x} - 1.96\widehat{\sigma}_x/\sqrt{N}, \, UB = \bar{x} + 1.96\widehat{\sigma}_x/\sqrt{N},$$

where

$$\bar{x} = \sum_{i=1}^{N} x_i / N$$

is the sample average and

$$\hat{\sigma} = \sqrt{\sum_{i=1}^{N} (x_i - \bar{x})^2 / N}$$

is the sample standard deviation.

Now we present a short review of working with so-called *indicator* random variables. Indicator random variable X has two values 1 and 0, $P(X = 1) = p$, $P(X = 0) = 1 - p = q$. Obviously, the average value of X,is $E[X] = p$. Suppose we have a family of indicator variables $X_1, X_2, \ldots, X_N$, $P(X_i = 1) = p_i$.

Then the average value of their linear combination is found by the following formula:

$$E[a_1 X_1 + a_2 X_2 + \cdots + a_N X_N] = \sum_{i=1}^{i=N} a_i p_i. \qquad (1.5.2)$$

This formula is correct also for the case that $X_i, i = 1, \ldots N$ are **not** independent random variables.

Suppose we have a *product* of two *independent* indicator random variables, $Y = X_i \cdot X_j$. Then

$$E[X_i \cdot X_j] = E[X_i]E[X_j] = p_i p_j \qquad (1.5.3)$$

This formula is **not** true if the indicator variables are not independent.

It is worth to make a small comment: the *product* of any number of indicator random variables is always an indicator random variable.

More information on probability and statistics and their applications in the engineering can be found in literature, e.g. in [3–5].

References

1. Samaniego, F. G. (2007). *System signatures and their applications in engineering reliability.* US: Springer.
2. Elperin, T., Gertsbakh, I. B., & Lomonosov, M. (1991). Estimation of network reliability using graph evolution models. *Transactions on Reliability, 40*(5), 572–581.
3. Bertsekas, D., & Tsitsliksis, J. (2008). *Introduction to probability* (2nd ed.). Belmont: Athena Scientific.
4. Ross S. M. (2019). *Introduction to probability models* (12th ed.). Cambridge: Academic Press.
5. Google: Basic statistics.

Chapter 2
Networks and Examples

Abstract This chapter starts with network definition, describes various types of network failure and presents an overview of various network failure criteria. For finite networks, most of these criteria are related to the loss of connectivity between special nodes called terminals. Classical network theory studies infinite network behaviour under node/edge failures, and the main object of interest is the so-called giant component of the network.(see, e.g. [1]) For finite networks, the analogue of it is the largest connected component or a largest cluster in a disintegrating network with failing nodes or edges. We present examples of networks where failure are related to the maximal cluster/component size. The second section is devoted to the simplest network structures–so-called series–parallel and parallel-series networks. We demonstrate that for independently failing elements, reliability analysis of these networks is rather simple.

Keywords Network *UP/DOWN* states · Terminal connectivity · Maximal component · Maximal cluster · Series/parallel systems

2.1 Nodes, Edges, Terminals

We meet networks every day and everywhere in our life. For formal study of network properties we must operate with abstract *models* of networks. In further, our principal network model will be a triple $\mathbf{N} = (V, E, T)$, where V is the *vertex* or *node* set, $|V| = m$, E is the *edge* or *link* set, $|E| = n$, and T is a set of special nodes called *terminals*, $T \subseteq V$, $|T| = h$ (Fig. 2.1).

In simple words, a network is a collection of circles (nodes) and links, i.e. line segments connecting the nodes. Terminals are marked as bold circles, like in Fig. 2.2

Our exposition will be centred around network behaviour when its elements (nodes and/or links) fail. We will deal with so-called *binary* elements which can be in two states *up* and *down* denoted by 1 and 0, respectively. When speaking about links, link i failure means that this link is erased, i.e. it does not exist. The state of link i, $i = 1, \ldots, n$ is denoted by binary variable x_i. If $x_i = 1$, link i is *up*; if $x_i = 0$, link i is *down*. x_i is often called link indicator variable.

© The Author(s), under exclusive license to Springer Nature Singapore Pte Ltd. 2020
I. Gertsbakh and Y. Shpungin, *Network Reliability*,
SpringerBriefs in Electrical and Computer Engineering,
https://doi.org/10.1007/978-981-15-1458-6_2

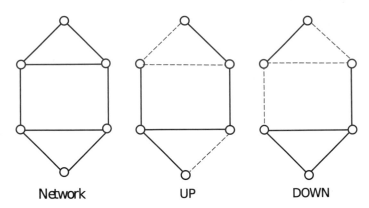

Network   UP   DOWN

Fig. 2.1 Network with unreliable edges. All–node connectivity criterion

In some models, the elements subject to failure are network *nodes* (vertices). If the indicator variable of node j is $y_j = 0$, i.e. node j is *down*, it means that all links incident to node j are *erased*, but the node itself remains intact. By an agreement, the terminals *do not fail*.

By *network state* we mean the set of all its elements (nodes and edges) that are in *up* state. We will distinguish network *UP* (operating) and *DOWN* (non-operating) states according to a certain criterion.

In many (but not all!) situations such criterion deals with the presence or absence of the *connection* between the terminals. We say that two nodes a and b are *connected* if there is a *path* of non-erased links connecting them.

Below we give examples of *UP* and *DOWN* states for topologically identical networks. The difference between these examples will be in different criteria and/or which elements (nodes or edges) are reliable.

Network with unreliable edges. All–node connectivity criterion
Let's have a look at Fig. 2.1. It shows three networks. The first is the network itself. It has 6 nodes and 8 edges and no terminals. Assume that the edges may be in *up* and *down* states, but the nodes are *reliable*, i.e. they are always in the *up* state. We say that the network is in the *UP* state if each pair of its nodes is connected by some path of non-erased edges. Such criterion is called *all–node connectivity*.

Initially the network is in the *UP* state. In the second network (in the middle) we see three erased edges (marked by red dashed line). Clearly that the network is in the *UP* state, by our criterion. The third shows a *DOWN* state of the network because not all pairs of nodes are connected by path of non-erased edges.

Network with unreliable edges. Three–terminal connectivity criterion
Figure 2.2 shows a network similar to the one on the left of Fig. 2.1, but with nodes a, b, c defined as terminals. Here the criterion is *three–terminal connectivity*, i.e. the network is *UP* if each pair of the terminals is connected by a path of non-erased edges. According to this definition, we see the *UP* state (in the middle). In the *DOWN* state, on the right, terminal a is isolated from two other terminals.

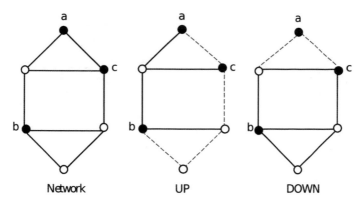

Fig. 2.2 Network with unreliable edges. Three terminal connectivity criterion

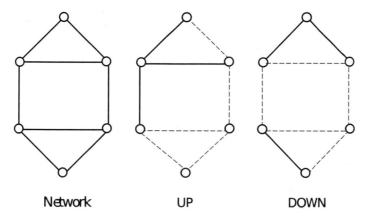

Fig. 2.3 Network with unreliable edges. Maximal component criterion

Network with unreliable edges. Maximal component criterion

For the next example we need the definition of a *component*. A subset $V_1 \subset V$ is called an *isolated component* of **N** if all nodes of V_1 are connected to each other and there are no edges of type $e = (a, b)$, where $a \in V_1$ and $b \in V - V_1$. An isolated node is considered as an isolated component. The *size* of a component is the number of nodes in it.

For example, in Fig. 2.1, in state *UP*, we see one component (all elements marked by black lines) of size 6, and in state *DOWN* there are two components, one of size 2 and the second of size 4.

Network in Fig. 2.3 has no terminals, edges are unreliable. The criterion here is the *maximal component size*. Namely, the network is in *UP* state if the maximal component has at least x nodes, where x is some given number. Suppose that $x = 4$. We see that the *UP* state has three components: one of size 4 and two of size 1 (isolated nodes). The *DOWN* state also has three components: one of size 3, one of size 2, and one of size 1.

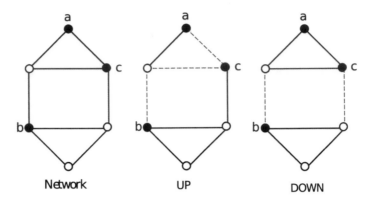

Fig. 2.4 Network with unreliable edges. Maximal cluster criterion

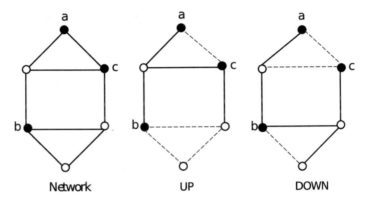

Fig. 2.5 Network with unreliable edges. Minimal path criterion

Network with unreliable edges. **Maximal cluster criterion**
The criterion in next example deals with the notion of a *cluster*.

By the definition, an isolated component of **N** is called a *cluster* if it contains at least one terminal node. A single terminal node is considered as a cluster of size 1.

Define now one more criterion. We say that the network is *UP* if it contains a cluster of size at least x, where x is some given number. Suppose that $x = 4$. In state *UP* on Fig. 2.4 there are two clusters, one of size 2 and the second of size 4. In *DOWN* state, we see two clusters, each of size 3.

Network with unreliable edges. **Minimal path criterion**
The following example deals with *minimal path criterion*. We say that the network is *UP* if each pair of terminals is connected by a path of length $\leq x$, where x is some given number. Suppose now that $x = 2$. We see in the *UP* state of Fig. 2.5 that for each pair of terminals there is a path (of non-erased edges!) of length 2. At the same time, in the *DOWN* state the only path between terminals a and c is of length 4.

In the following examples we will consider networks with *unreliable nodes*. We remind that if some node is *down*, then all edges incident to it are erased, and the

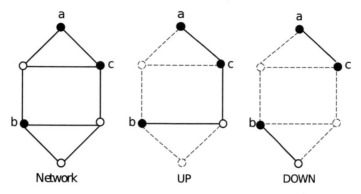

Fig. 2.6 Network with unreliable nodes. Terminal Connectivity

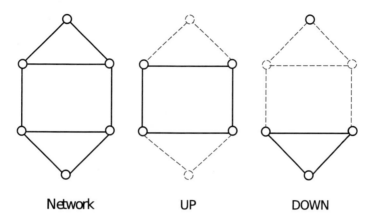

Fig. 2.7 Network with unreliable nodes. Maximal component criterion

node itself remains intact. In this case the node and the appropriate edges will be marked by red dashed lines.

Network with unreliable nodes. Three terminal connectivity
The network on Fig. 2.6 has three terminals and the *UP* criterion is terminal connectivity. We see that in the *UP* state two nodes are *down*, but all three terminals are connected by non-erased edges. In the case of *DOWN* the network splits into two isolated clusters.

Network with unreliable nodes. Maximal component criterion
Figure 2.7 shows network without terminals and with maximal component criterion. Suppose that $x = 4$. We see that in the *UP* state, there is component of size 4, while the *DOWN* state contains one component of size 3 and one of size 1.

Network with unreliable nodes. Maximal cluster criterion
The network in Fig. 2.8 has **two** terminals. The criterion is the size of maximal cluster. Suppose that $x = 3$. We see that the *UP* state has two clusters. One of size 3 and

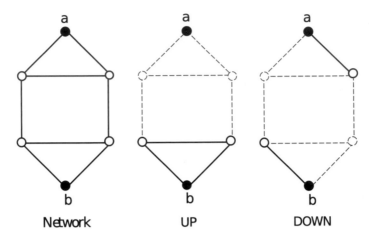

Fig. 2.8 Network with unreliable nodes. Maximal cluster criterion

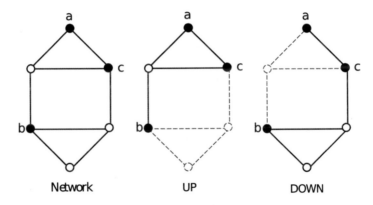

Fig. 2.9 Network with unreliable nodes. Minimal path criterion

another of size 1. In the *DOWN* state, the network has disintegrated into two clusters, both of size 2.

Network with unreliable nodes. Minimal path criterion
The criterion for network in Fig. 2.9 is the length of minimal path between terminals. Suppose that $x = 2$. In the *UP* state we see a path of length 1 between a and c, and paths of length 2 between a and b, and b and c. In the *DOWN* state, the minimal path between a and b is of size 3.

So far only undirected networks have been considered. The following example deals with an *oriented* network.

Oriented flow network with unreliable edges. Maximal flow criterion
In an *oriented network*, the direction of the flow along the edge is determined by an arrow. In *flow* network, for each edge $e = (x, y)$ directed from node x to node y, we define the maximal flow $c(e)$, where c stands for flow *capacity* which can be

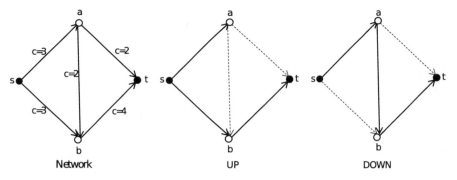

Fig. 2.10 Flow Network with unreliable edges

delivered from x to y along this edge. We say that the network state is in UP if the maximal flow from source to sink is not less than some given value Φ, and in state $DOWN$, otherwise.

Figure 2.10 represents oriented flow network with 4 nodes and 5 edges. The node s is the source and t is the sink. The edge capacities are as follows: $c(s, a) = 3$, $c(s, b) = 3$, $c(a, t) = 2$, $c(a, b) = 2$, $c(b, t) = 4$. It is easy to check that the maximal flow from source s to sink t equals 6. For example, it may be obtained by the following flows $w(i, j)$: $w(s, a) = 3$, $w(a, t) = 2$, $w(a, b) = 1$, $w(s, b) = 3$, $w(b, t) = 3$. Suppose that we define the UP state for this network as a state with maximal flow not less than $\Phi = 3$. In Fig. 2.10 UP, edges (a, t) and (a, b) are *down*, and the maximal flow equals 3. In $DOWN$, edges (a, t) and (s, b) are *down*, and the maximal flow is 2.

2.2 Series-Parallel Networks

Probably, the simplest networks are networks having series-parallel and parallel–series structure. Figure 2.11a presents a *series* network having terminals s and t, and four edges $1 = (s, a)$, $2 = (a, b)$, $3 = (b, c)$ and $4 = (c, t)$. Elements subject to failure are the edges. Failure of an edge is loss of connection between its endpoints. Edge i has *up* probability p_i and *down* probability $q_i = 1 - p_i$. It will be assumed that the edges fail independently. By definition, the series network is UP if all edges are *up*, which provides the $s - t$ connection. The failure of any edge leads to $DOWN$ of the network. A natural example of a series network might be a road between cities s and t, having four bridges subject to failure. $s - t$ connection is disrupted if at least one of the bridges is *down*. It is easy to find the probability that the series network is UP:

$$P_{ser}(UP) = p_1 \cdot p_2 \cdot p_3 \cdot p_4. \qquad (2.2.1)$$

Figure 2.11b presents a *parallel* network with 5 edges e_1, e_2, e_3, e_4, e_5 each connecting nodes s and t. This network fails if the connection between s and t is disrupted,

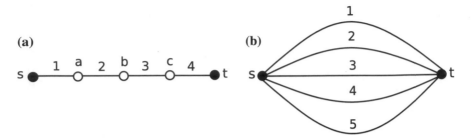

Fig. 2.11 Series (**a**) and Parallel (**b**) network

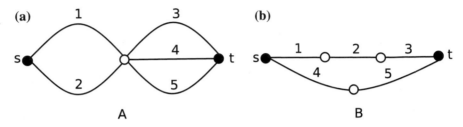

Fig. 2.12 Series-parallel (**a**) and Parallel-series (**b**) network

and this happens if all edges fail. Probability of this event (assuming independence) is

$$P_{par}(DOWN) = q_1 \cdot q_2 \cdot q_3 \cdot q_4 \cdot q_5. \tag{2.2.2}$$

Probably the readers already have in mind the possibility to join several series networks in parallel connection and several parallel networks in series connection, see Fig. 2.12a, b. A shows a series connection of two parallel networks with two and three edges in parallel. B shows a parallel connection of two series networks having three and two edges. Let us compute the UP probabilities for these two networks. Again the edges fail independently and edge e_i is up with probability p_i. First we note that

$$P_A(UP) = P(left - UP) \cdot P(right - UP) = [1 - q_1 \cdot q_2] \cdot [1 - q_3 \cdot q_4 \cdot q_5] \tag{2.2.3}$$

Similarly,

$$P_B(DOWN) = P(upper - DOWN) \cdot P(lower - DOWN)$$
$$= [1 - p_1 p_2 p_3][1 - p_4 p_5], \tag{2.2.4}$$

and

$$P_B(UP) = 1 - P_B(DOWN). \tag{2.2.5}$$

It is worth mentioning that series-parallel networks are widely used in many technical areas, for example, in computing system reliability. Suppose that it is necessary to maintain low temperature in a chemical laboratory, and for that purpose three operating air conditioners are needed. Failure of any one of them means system failure. This is an example of a series system with three components. Another example: water supply system from water source s to high buildings in remote city t. This system has three water pumping stations located between s and t_i. In city, the water pipe splits into five smaller water pipes delivering the water to buildings $t_1, t_2, \ldots, t_5$. Each t_i has his own water pump. System failure, by definition, happens if one of t_i buildings remains without water. So, here the pumping stations are connected in series and the water pumps–in parallel and the whole system is a series-parallel network. Series-parallel systems consisting of independent elements have a valuable property: their UP/DOWN probabilities are easily computed if the element failure probabilities are known.

In supply systems, very popular is a modification of a parallel system called $K - out - of - N$ system. It is a network with one source terminal s and N t-type terminals $t_1, t_2, \ldots, t_N$. By the definition, the system is UP if at least K out of N of these terminals are connected to the source s. Returning to the above water supply system, let $N = 4$ and $K = 2$, i.e. the system is operational (i.e. is UP) if at *least* two out of four buildings get water supply. Suppose that the probability that a building has water supply is p_0. Then the number X of buildings receiving water supply is a binomial random variable

$$X \sim \mathbf{B}(4, p_0).$$

In our example,

$$P(UP) = P(X \geq 2) = 6p_0^2(1 - p_0)^2 + 4p_0^3(1 - p_0) + p_0^4.$$

More information on networks and reliability can be found in the literature, e.g. in [2–5].

References

1. Newman, M. E. J. (2010). *Networks: an introduction*. Oxford: Oxford University Press.
2. Barlow, R. E., & Proschan, F. (1981). *Statistical theory of reliability and life testing*. To Begin With; Reprint Edition.
3. Gertsbakh, I. (2000). *Reliability theory with applications to preventive maintenance*. Berlin: Springer.
4. Gertsbakh, I., & Shpungin, Y. (2009). *Models of network reliability: analysis, combinatorics and Monte Carlo*. Boca Raton: CRC Press.
5. Gertsbakh, I., & Shpungin, Y. (2011). *Network reliability and resilience*. Berlin: Springer.

Chapter 3
Direct Network Reliability Calculation

Abstract This chapter is a collection of several direct methods for calculating network reliability. These methods are good for small networks having not more than 8–10 elements subject to failure. The most straightforward is network state enumeration method. It allows obtaining network $UP/DOWN$ probability in an explicit form. More elegant and efficient are the methods based on using network structural parameters—the so-called minimal *path* and minimal *cut* sets. Knowing these parameters allows us to construct rather accurate upper and lower bounds on network reliability. For very reliable networks in which node/edge failure reliability tends to zero, very good results are obtained by using so-called Burtin-Pittel approximation. We present it in Sect. 3.5. Network reliability as a function of the number of failed elements is an important characteristic of so-called *resilience*, see Sect. 3.6. The chapter is concluded by introducing so-called *dynamic reliability* which is network reliability behaviour in time. This is obtained by replacing element static reliability p by survival probability $1 - G(t)$ depending on time t.

Keywords Min cuts · Min paths · Reliability bounds · B-P approximation · Network resilience · Dynamic reliability

3.1 Static Reliability and Structure Function

In this chapter we will consider methods of calculating network reliability in a direct way, without using Monte Carlo. To make these calculations more efficient we need to introduce some new notions and definitions. We consider only networks with *binary* elements. Each binary element can be in two states: *up* (operational) and *down* (failure). A compact description of all network element states will be given by the vector

$$\mathbf{x} = (x_1, x_2, \ldots, x_n),$$

where $x_i = 1$ if element i is *up* and $x_i = 0$ if this element is *down*. The networks are assumed to be also in two states—*UP* and *DOWN*. Network state is described by so-called *structure function* $\phi(\mathbf{x})$.

© The Author(s), under exclusive license to Springer Nature Singapore Pte Ltd. 2020
I. Gertsbakh and Y. Shpungin, *Network Reliability*,
SpringerBriefs in Electrical and Computer Engineering,
https://doi.org/10.1007/978-981-15-1458-6_3

Fig. 3.1 4 edge and 4 node network is *UP* if terminals *s* and *t* are connected

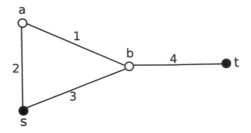

Definition 3.1 *Structure function* $\phi(\mathbf{x})$ ([1, 2])
1. $\phi(\mathbf{x}) = 0$ if the network state $\mathbf{x}$ is *DOWN*.
2. $\phi(\mathbf{x}) = 1$ if the network state $\mathbf{x}$ is *UP*.

Remark 3.1 From the definition of $\phi(\mathbf{x})$ it follows that network reliability equals $R = P(\phi(\mathbf{x}) = 1) = E[\phi(\mathbf{x})]$.

Remark 3.2 We consider only monotone networks, which means the following.
1. $\phi(0, 0, \ldots, 0) = 0$.
2. $\phi(1, 1, \ldots, 1) = 1$.
3. Let $\mathbf{u}$ and $\mathbf{v}$ be network states such that $\mathbf{u} \leq \mathbf{v}$. Then $\phi(\mathbf{u}) \leq \phi(\mathbf{v})$. In words: the network state cannot turn from *UP* to *DOWN* if an element changes its state from *down* to *up*. This condition may seem redundant, because it is satisfied for all the *UP* criteria **we are considering**. However, in some rare cases this is not so.

Suppose now that element i is *up* with probability p_i and *down* with probability $q_i = 1 - p_i$. **All elements are assumed to be statistically independent**. Denote by $U(\mathbf{x})$ and $D(\mathbf{x})$ the sets of indices corresponding to the *up*-elements of $\mathbf{x}$ and *down*-elements of $\mathbf{x}$ respectively. Then the probability that the network is in state $\mathbf{x}$ equals to

$$P(\mathbf{x}) = \prod_{i \in U(\mathbf{x})} p_i \cdot \prod_{j \in D(\mathbf{x})} q_j \tag{3.1.1}$$

For example, if $\mathbf{x} = (0, 1, 1, 0, 0, 1, 1, 1)$ then $P(\mathbf{x}) = p_2 p_3 p_6 p_7 p_8 q_1 q_4 q_5$. Finally, the *static reliability* of network is defined as $R(N) = P(N \text{ is } UP)$ and we have

$$R(N) = \sum_{\mathbf{x} \in UP} P(\mathbf{x}) = \sum_{\mathbf{x} \in UP} \left(\prod_{i \in U(\mathbf{x})} p_i \cdot \prod_{j \in D(\mathbf{x})} q_j \right) \tag{3.1.2}$$

Let us consider an example of computing *UP* probability of a small network shown on Fig. 3.1.

This network is *UP* if the nodes s and t are connected. The network has 4 edges. Each of them is binary, and edge $i, i = 1, 2, 3, 4$ is *up* with probability p_i and *down*

Table 3.1 State vectors of network on Fig. 3.1

y	x_1	x_2	x_3	x_4	$\phi(\mathbf{y})$	$P(\mathbf{y})$
$\mathbf{y}(1)$	0	0	0	0	0	$q_1 q_2 q_3 q_4$
$\mathbf{y}(2)$	0	0	0	1	0	$q_1 q_2 q_3 p_4$
$\mathbf{y}(3)$	0	0	1	0	0	$q_1 q_2 p_3 q_4$
$\mathbf{y}(4)$	0	0	1	1	1	$q_1 p_2 p_3 p_4$
$\mathbf{y}(5)$	0	1	0	0	0	$q_1 p_2 q_3 q_4$
$\mathbf{y}(6)$	0	1	0	1	0	$q_1 p_2 q_3 4_4$
$\mathbf{y}(7)$	0	1	1	0	0	$q_1 p_2 p_3 q_4$
$\mathbf{y}(8)$	0	1	1	1	1	$q_1 p_2 p_3 p_4$
$\mathbf{y}(9)$	1	0	0	0	0	$p_1 q_2 p_3 q_4$
$\mathbf{y}(10)$	1	0	0	1	0	$p_1 p_2 p_3 p_4$
$\mathbf{y}(11)$	1	0	1	0	0	$q_1 p_2 p_3 q_4$
$\mathbf{y}(12)$	1	0	1	1	1	$p_1 q_2 p_3 p_4$
$\mathbf{y}(13)$	1	1	0	0	0	$q_1 p_2 p_3 q_4$
$\mathbf{y}(14)$	1	1	0	1	1	$p_1 q_2 p_3 p_4$
$\mathbf{y}(15)$	1	1	1	0	0	$p_1 p_2 p_3 q_4$
$\mathbf{y}(16)$	1	1	1	1	1	$p_1 p_2 p_3 p_4$

with probability $q_i = 1 - p_i$. Edges are independent. The total number of network states is $2^4 = 16$. All these states are shown in the Table 3.1. Vector $\mathbf{y}(i)$ describes the edge states. $P(\mathbf{y})$ is the probability that the edges of the network are in state $\mathbf{y}$. The column $\phi(\mathbf{y})$ shows the network *UP/DOWN* state. It is easy to see that the network is *UP* if edges 4 and 3 are *up*, or 4 is *up* and either all other edges are *up*, or any two of three edges 1, 2, 3 are *up*. In total, there are five *UP* states. The probability that s and t are connected equals

$$P(UP) = q_1 p_2 p_3 p_4 + p_1 q_2 p_3 p_4 + p_1 p_2 q_3 p_4 + q_1 q_2 p_3 p_4 + p_1 p_2 p_3 p_4. \quad (3.1.3)$$

This method (all state enumeration) works good for very small networks having 4–6 edges. If the number of edges is $n = 8 - 20$ you need a computer to make the list of all 2^n states and for finding the network *UP* states.

An alternative calculations scheme can be used if we know network structural parameters, so-called network *minimal path* sets and/or *minimal cut* sets.

Definition 3.2 (*Cut vector, minimal cut set* [1, 2])
A state vector $\mathbf{x} = (x_1, x_2, \ldots x_n)$ is called *cut vector* if $\phi(\mathbf{x}) = 0$.

The set $C(\mathbf{x}) = \{i : x_i = 0\}$ is called *minimal cut set* or *minimal cut* if there is no such element $j \in C(\mathbf{x})$ that after its removal $C(x) - x_j$ is no more a cut set.#

For example, in bridge network (Fig. 3.2), the set $(0, 0, 0, 1, 1)$ is a cut vector. The set $(x_1 = 0, x_2 = 0, x_3 = 0)$ is a cut set, but not a minimal cut set. If we remove edge 3, the set $(x_1 = 0, x_2 = 2)$ remains a cut set. Clearly, it is now a *minimal cut set*.

Fig. 3.2 *s-t* bridge network

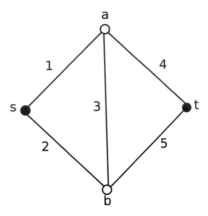

Definition 3.3 *Path vector, minimal path set* ([1, 2])
A state vector $\mathbf{x} = (x_1, x_2, \ldots x_n)$ is called *path vector* if $\phi(\mathbf{x}) = 1$.
The set $T(\mathbf{x}) = \{i : x_i = 1\}$ is called *minimal path set* or *minimal path* if there is no
such element $j \in T(\mathbf{x})$ that after its removal $T(x) - x_j$ is no more a path set.#
For example, in bridge network, $(1, 0, 1, 1, 0)$ is a path vector and $(x_1 = 1, x_3 = 1, x_4 = 1)$ is a path set, but not a minimal path set because after removal of edge 4,
$(x_1 = 1, x_3 = 1)$ remains a path set and is a *minimal* path.

Suppose we know all m minimal paths $T_1, T_2, \ldots, T_m$ and all k minimal cuts
$C_1, C_2, \ldots, C_k$ of the network. Then we have the following

Theorem 3.1

$$\phi(\mathbf{x}) = 1 - \prod_{j=1}^{m}\left(1 - \prod_{i \in T_j} x_i\right). \tag{3.1.4}$$

$$\phi(\mathbf{x}) = \prod_{j=1}^{k}\left(1 - \prod_{i \in C_j}(1 - x_i)\right). \tag{3.1.5}$$

We will omit a simple proof of this theorem.

Theorem 3.1 gives us the possibility to obtain an expression for bridge reliability
R_0. Let us define a binary *random variable* X_i for edge i of the bridge. We will assume
that $X_i = 1$ if edge i is *up* and $X_i = 0$, otherwise. Assume also that bridge edges are
independent. According to Theorem 3.1 we will obtain the following expression for
$\phi(\mathbf{X})$:

$$\phi(\mathbf{X}) = 1 - (1 - X_1 X_4)(1 - X_2 X_5)(1 - X_1 X_3 X_5)(1 - X_2 X_3 X_4) \tag{3.1.6}$$

because $(1, 4), (2, 5), (1, 3, 5)$ and $(2, 3, 4)$ are minimal paths of the bridge. Let us
assume that all $p_i = p$ and open the brackets. Note that $X_i^r = X_i$ for each $r > 0$,
and by independence $E[X_i \cdot X_j] = E(X_i) \cdot E(X_j)$. After some algebra, we obtain
the following formula for $E[\phi(\mathbf{X})] = P(\phi(X) = 1)$:

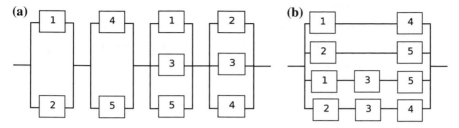

Fig. 3.3 Series connection of parallel min cuts (**a**) and parallel connection of series min paths (**b**) for bridge network on Fig. 3.2

$$E[\phi(X)] = R_0 = 2p^2 + 2p^3 - 5p^4 + 2p^5. \tag{3.1.7}$$

Figure 3.3 represents a series connection of four min cuts of the bridge and a parallel connection of its four min paths. In the picture on the left and on the right each element appears several times and we conclude that systems comprising series and parallel systems are NOT independent. But it turns out that if the elements on the left and the right with identical numbers are *independent* then the reliability of series-parallel system on the left and parallel-series system on the right produce lower and upper bounds, respectively, on the true system reliability R_0 (Fig. 3.4).

An interesting and important fact is that the system on the left (series connection of four parallel systems) is less reliable than the bridge, i.e. produces lower bound on R_0. In addition, the system on the right (parallel connection of series systems) is always more reliable than the bridge system. Let us use these facts to try to check how good are these bounds on true reliability R_0. For this purpose we need to find out the reliability of the "lower" and "upper" systems.

Parallel system of two identical elements fails if both elements fail and remains *UP* with probability $1 - q^2$. The "left" system remains *UP* if all four subsystems are *up*. This leads us to the expression

$$LB(q) = (1 - q^2)^2 (1 - q^3)^2$$

where $q = 1 - p$. The upper bound is a parallel connection of four series system. It is easy to obtain that

$$UB(p) = 1 - (1 - p^2)^2 (1 - p^3)^2.$$

3.2 Counting Paths and Cuts

Let us return to Table 3.1 which presents all 2^4 binary states of our network on Fig. 3.1. We see that the network has 5 *UP* states and the expression (3.1.1) shows the structure of these states. This fact will be evident if we replace p_i by p and q_i by q. Then we obtain that

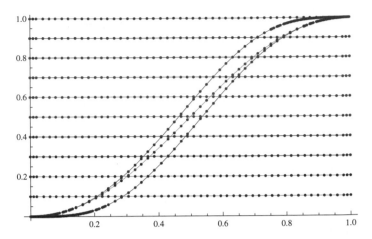

Fig. 3.4 Upper and lower bounds on the exact reliability function for the bridge network

$$P(UP) = 3qp^3 + q^2p^2 + p^4. \tag{3.2.1}$$

From here we make an important observation that there are 3 path vectors with one element *down* and 3 elements *up*, 1 path vector with 2 elements *up* and two *down*, and one path vector with all 4 elements *up*. All these facts we can easily check on Fig. 3.1.

Let us check the situation with the cut vectors. Each state vector **x** is either a path vector or a cut vector. The Table 3.1 shows that there are 5 path vectors out of 16, and therefore there are 11 cut vectors. It is interesting to see the structure of these vectors. The simplest way to do this, is the following. Note that $P(DOWN) + P(UP) = 1$; replace 1 by $(p + q)^4$ and simplify the resulting expression:

$$P(DOWN) = 1 - P(UP) = (p + q)^4 - 3qp^3 - q^2p^2 - p^4 =$$
$$qp^3 + 5q^2p^2 + 4q^3p + q^4. \tag{3.2.2}$$

Now it is obvious that there are 5 cut vectors with two elements *down*, 4 cut vectors with 3 elements down, one cut vector with four elements *down* and one cut vector with one element *down*. The reader will easily identify the cut vectors from Fig. 3.1.

Remark 3.3 We will return to this example in Chap. 6 when we introduce the so-called structural invariant—destruction spectrum. Now we make only one important observation: the number of cut and path vectors (and therefore, the cut and path sets) of any size *does not* depend of p or q values.#

Remark 3.4 Let $C(k)$ be the number of *cut* sets of size k in the network with n elements. Let $W(k) = C(k)q^k p^{(n-k)}$ be the "probability mass" of this cut set. Obviously, $A = \sum_{k=1}^{n} W(k) = P(DOWN)$. Similarly, let $T(k)$ be the number of *path* sets of size k in the network. Let $V(k) = T(k)p^k q^{(n-k)}$ be the "probability mass" of this path set. Obviously, $B = \sum_{k=1}^{n} V(k) = P(UP)$. These sums together equal 1 and it can be represented as $(p+q)^n$. Here the coefficient at $q^k p^{n-k}$ is $n!/(k!(n-k)!)$. It must be equal to the similar coefficient in the sum of A and B. We obtain, therefore, that $C(k) + T(n-k) = n!/(k!(n-k)!)$.#

3.3 Using Inclusion/Exclusion Formula

In this section we give some examples of computing network reliability using inclusion/exclusion formula.

Theorem 3.2 *Let* $T_1, T_2, \ldots, T_s$ *be all minimal paths of the network and* $C_1, C_2, \ldots,$ C_k *be all minimal cuts of the network. Let now* A_i *be the following event:* $A_i = \{All$ *elements of the path* T_i *are in the* up *state}. Define* $B_i = \{All$ *elements of the cut* C_i *are in the state* down}. *Then network reliability R can be represented as*

$$R = P\left(\bigcup_{i=1}^{s} A_i\right) \tag{3.3.1}$$

or

$$R = 1 - P\left(\bigcup_{i=1}^{k} B_i\right) \tag{3.3.2}$$

We will omit a simple proof of this theorem.

The appropriate calculations in (3.3.1) and (3.3.2) can be performed using the following *inclusion/exclusion* formula. Let $D_1, \ldots, D_n$ be some set of events. Then

$$P\left(\bigcup_{i=1}^{n} D_i\right) = \sum_{i=1}^{n} P(D_i) - \sum_{i<j} P(D_i \cap D_j)$$

$$+ \sum_{i<j<k} P(D_i \cap D_j \cap D_k) - \cdots + (-1)^{n-1} \sum_{i<\cdots<n} P\left(\bigcap_{i=1}^{n} D_i\right), \tag{3.3.3}$$

Below we present some examples of using this formula for computing network reliability.

Network with Non-reliable Edges. Terminal Connectivity Criterion

Consider the bridge on Fig. 3.2. It is easy to check that the minimal cuts are: $(1, 2)$, $(4, 5)$, $(1, 3, 5)$, $(2, 3, 4)$. The minimal paths are: $(1, 4)$, $(2, 5)$, $(1, 3, 5)$, $(2, 3, 4)$. Since the number of all minimal paths and all minimal cuts is the same, it does not matter how to calculate bridge reliability: using minimal cuts or minimal paths. We will calculate using minimal paths.

Denote $A_i =\{$all elements in T_i are in $up\}$. Then $R = P(A_1 \cup A_2 \cup A_3 \cup A_4) = P(A_1) + P(A_2) + P(A_3) + P(A_4) - P(A_1 \cap A_2) - P(A_1 \cap A_3) - P(A_1 \cap A_4) - P(A_2 \cap A_3) - P(A_2 \cap A_4) - P(A_3 \cap A_4) + P(A_1 \cap A_2 \cap A_3) + P(A_1 \cap A_2 \cap A_4) + P(A_1 \cap A_3 \cap A_4) + P(A_2 \cap A_3 \cap A_4) - P(A_1 \cap A_2 \cap A_3 \cap A_4).$

Let us assume that all $p_i = p$. Then we obtain: $R = 2p^2 + 2p^3 - 5p^4 + 2p^5.$

Network with Non-reliable edges. Maximal cluster criterion

Consider again the bridge in Fig. 3.2. Suppose that $x = 3$, i.e. the network is UP if its maximal cluster is at least of size 3.

Let us find first the minimal paths. From the bridge figure we see that each pair of *adjacent* edges forms a cluster of size 3, so each such pair is a minimal path. There are the following pairs: $T_1 = (1, 2), T_2 = (1, 3), T_3 = (1, 4), T_4 = (2, 3), T_5 = (2, 5), T_6 = (3, 4), T_7 = (3, 5), T_8 = (4, 5)$. We have 8 minimal paths for such small network!

So, **manual calculation** of network reliability using 3.3.1 or structure function is quite cumbersome.

Let's find all minimal cuts. Obviously, they are: $C_1 = (1, 3, 5), C_2 = (2, 3, 4)$ and $C_3 = (1, 2, 4, 5)$. Denote $B_i =\{$all elements of C_i are in $down\}$. We get $R = 1 - P(B_1 \cup B_2 \cup B_3) = 1 - P(B_1) - P(B_2) - P(B_2) + P(B_1 \cap B_2) + P(B_1 \cap B_3) + P(B_2 \cap B_3) - P(B_1 \cap B_2 \cap B_3) = 1 - q_1q_3q_5 - q_2q_3q_4 - q_1q_2q_4q_5 + q_1q_2q_3q_4q_5 + q_1q_2q_3q_4q_5 + q_1q_2q_3q_4q_5 - q_1q_2q_3q_4q_5.$

For the case $q_i \equiv q$ we get $R = 1 - 2q^3 - q^4 + 2q^5.$

Network with Non-reliable nodes. Maximal cluster criterion

The network on Fig. 3.5 has two terminals–s and t. The criterion is the size of the maximal cluster. Suppose that $x = 4$, i.e. the network is UP if there is cluster of size at least 4.

Right from the figure we see that all minimal paths are the following pairs: $T_1 = (a, d), T_2 = (b, c), T_3 = (b, d), T_4 = (c, d)$. Denote $A_i = \{$all elements in T_i are in $up\}$. Let us calculate network reliability using 3.3.1 and 3.3.3.

$R = P(A_1 \cup A_2 \cup A_3 \cup A_4) = P(A_1) + P(A_2) + P(A_3) + P(A_4) - P(A_1 \cap A_2) - P(A_1 \cap A_3) - P(A_1 \cap A_4) - P(A_2 \cap A_3) - P(A_2 \cap A_4) - P(A_3 \cap A_4) + P(A_1 \cap A_2 \cap A_3) + P(A_1 \cap A_2 \cap A_4) + P(A_1 \cap A_3 \cap A_4) + A_2 \cap A_3 \cap A_4) - P(A_1 \cap A_2 \cap A_3 \cap A_4).$

Suppose that $p_i \equiv p$ for all i. Then we get: $R = 4p^2 - p^4 - 5p^3 + 3p^4 + p^3 - p^4 = 4p^2 + p^4 - 4p^3.$

The minimal cuts of the network are: $C_1 = (b, d), C_2 = (c, d), C_3 = (a, b, c)$. Let $B_i =\{$all elements in C_i are in $down\}$. We get $1 - R = P(B_1 \cup B_2 \cup B_3) = P(B_1) + P(B_2) + P(B_3) - P(B_1 \cap B_2) - P(B_1 \cap B_3) - P(B_2 \cap B_3) + P(B_1 \cap B_2 \cap B_3).$

Suppose that $q_i \equiv q$ for all i. Then $1 - R = q^2 + q^2 + q^3 - q^3 - q^4 - q^4 + q^4$ and $R = 1 - 2q^2 + q^4.$ It is easy to check that in both cases we get the same answer.

Fig. 3.5 Network with
unreliable nodes. Maximal
cluster criterion

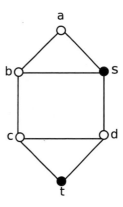

3.4 Pivotal Formula

Now we present a formula which will be very useful in our future exposition. It is called *pivotal* formula. First we note that if the network elements are binary, fail independently and element $i = 1, \ldots n$ is *up* with probability p_i, then network reliability R_0 can be represented as an average value of the structure function $\phi(X_1, \ldots, X_n)$:

$$R_0 = E[\phi(X_1, X_2, \ldots, X_n)] = \Psi(p_1, p_2, \ldots p_n).$$

Denote by $\Psi(p_1, p_2, \ldots, p_j = 1, \ldots, p_n)$ the reliability of our network when the jth element is permanently *up*, and by $\Psi(p_1, p_2, \ldots, p_j = 0, \ldots, p_n)$ the reliability of our network when the jth element is permanently *down*. Then the following theorem is true:

Theorem 3.3 Pivotal decomposition

$$R_0 = p_j \cdot \Psi(p_1, p_2, \ldots, 1_j = 1, \ldots, p_n) + (1 - p_j) \cdot \Psi(p_1, p_2, \ldots, 0_j, \ldots, p_n).$$
$$(3.4.1)$$

The proof is elementary and is based on the Law of Total Probability, see Chap. 1. The use of pivotal formula allows simplify considerably direct computation of network reliability, see e.g. [1]. In these lectures we will use the pivotal formula for finding the partial derivatives of network reliability function.

3.5 Burtin-Pittel Approximation

Let us consider in this section the case of a very reliable networks built of identical and independent elements. This means, in particular, that all elements have the same *down* probability q and *up* probability $1 - q$. "Very reliable" means that q is small, say of magnitude 0.01–0.1. Let us return to the expression (3.2.2) for network *DOWN* probability:

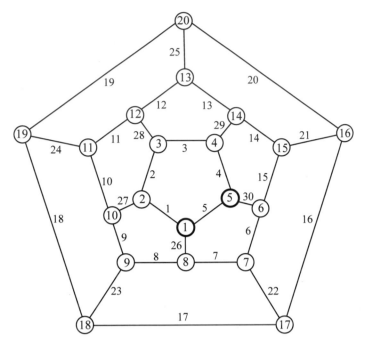

Fig. 3.6 Dodecahedron with 20 nodes and 30 edges. Edges fail independently. Nodes 1 and 5 are terminals

$$P(DOWN) = qp^3 + 5q^2p^2 + 4q^3p + q^4.$$

and rewrite it substituting $p = 1 - q$. After little algebra we obtain the expression

$$P(DOWN) = q + 2q^2 - 3q^3 + q^4,$$

If q is "small", formally $q \to 0$, then the last expression can be rewritten as

$$P(DOWN) = q(1 + O(q)) = q + O(q^2) \tag{3.5.1}$$

It means that $P(DOWN)$ is equal q up to a term which is of smaller magnitude than q. In physical terms, the *DOWN* probability depends mainly on the cut set of minimal size 1. What happens in this example is typical for very reliable networks: their *DOWN* probability is mainly determined by the cut sets of *minimal size*. This approximation was suggested by Burtin and Pittel in a slightly different form, see [1].

Let us consider the network shown in Fig. 3.6.

It is easy to see from Fig. 3.6 that there are two minimal cuts of size 3 each of which separates the terminals: (1, 5, 26) and (4, 5, 30). Therefore Burtin-Pittel approximation is $2q^3$. From Table 3.2 below we see that for $q < 0.05$ this approxi-

Table 3.2 Comparison of exact and approximate $DOWN$ probabilities

q	$P(DOWN)$–exact	$P_{approx}(DOWN) =$ $2q^3$	$rel.err.\%$
0.20	0.0362	0.0219	39
0.10	0.00288	0.00234	18
0.04	0.000146	0.000136	7
0.02	1.2×20^{-5}	1.65×10^{-5}	3
0.01	2.03×10^{-6}	2×10^{-6}	1.5

mation provides small relative error and might serve as a reasonable approximation to the true dodecahedron $DOWN$ probability.

Example *The lower bound to bridge reliability.*

Let us return to the lower bound of bridge reliability, see Fig. 3.3a which is a series connection of four parallel systems—two having two elements and two—three elements. Clearly this system has four minimal cuts: two having two elements each and two—having three elements. Burtin-Pittel approximation suggests to take the subsystem having two min cuts of two elements as an approximation to the true $DOWN$ probability of the lower bound. So, let Q_1 be the $DOWN$ probability of the system having two min cuts of two elements each. Obviously,

$$Q_1 = 1 - (1 - q^2)^2$$

Let Q_2 be the $DOWN$ probability of the system with four min cuts:

$$Q_2 = 1 - (1 - q^2)^2(1 - q^3)^2$$

Let us calculate the relative error in % in replacing Q_2 by Q_1.

$$rel.err. = 100 \cdot \frac{Q_2 - Q_1}{Q_1}. \qquad (3.5.2)$$

For $q = 0.1$ we obtain a reasonably small relative error of 10%.

3.6 Network Resilience

One of important concepts in network analysis is its *resilience* which is closely related to network reliability. In Chap. 5, we will give an efficient method to calculate resilience along with reliability, and in Chap. 8, we will give relevant examples for fairly large networks.

Wikipedia gives the following definition of resilience:

Resilience is considered as a superset to *survivability* which is defined as "capability to fulfil network mission in the presence of attacking, failures and accidents of network elements".

We would like to present here a more formal version of network *probabilistic resilience* by citing the definitions from [3]:

In the case of element random failures, the *disconnection probability* of a network **N** is defined as

$$P(\mathbf{N}; i) = P[\ \mathbf{N} \text{ is disconnected exactly after } i\text{-th failure}].$$

Another important notion is presented in the following

Definition 3.4 *Probabilistic resilience* [4]:

Let **N** be a network with n elements. The probabilistic resilience $res_{pr}(\mathbf{N}; \beta)$ is the largest number of element failures such that **N** is still *UP* with probability $1 - \beta$. Formally,

$$res_{pr}(\mathbf{N}, \beta) = max \left\{ I : \sum_{i=1}^{I} P(\mathbf{N}, i) \le \beta \right\}.\#$$

The authors of this definition assume that network element failures appear in *random order*, i.e. all $n!$ orderings are equally probable.

Let us demonstrate resilience computation for a bridge network on Fig. 3.2. The bridge failure is defined as disconnection of s and t. The bridge never fails when only one link is down. Therefore, $P(\mathbf{N}; 1) = 0$. We could analyse $5! = 120$ permutations of 5 edges to get all values of $P(\mathbf{N}; i)$, $i = 2, 3, 4, 5$, but in our simple case there is no need for this. Indeed, the network never fails after exactly one edge failure. The network fails after two edge failures only in the case of failure of the minimal cuts of size 2: $(1, 2), (4, 5)$. The number of permutations having in the first two positions one of these cuts is equal to $2!2!3! = 24$. So we get $P(\mathbf{N}; 2) = \frac{24}{120} = 0.2$. Consider the network failure at step 4. This can happen only if the first three positions contain all possible combinations of $(1, 3, 4)$ or $(2, 3, 5)$, and in the next positions the combination of the two remaining elements. So we get $P(\mathbf{N}; 4) = \frac{2 \cdot 3! 2!}{5!} = 0.2$. Clearly that $P(\mathbf{N}; 5) = 0$, and therefore we get $P(\mathbf{N}; 3) = 1 - 0.2 - 0.2 = 0.6$.

Now let us consider a "modified bridge" on Fig. 3.7.

For this network, in the same way as in the previous example, it is easy to get: $P(\mathbf{N}; 1) = 0$, $P(\mathbf{N}; 2) = 0$, $P(\mathbf{N}; 3) = 0.4$, $P(\mathbf{N}; 4) = 0.4$, $P(\mathbf{N}; 5) = 0.2$.

Following Definition 3.4 we obtain the following results:

For bridge: $res_{pr}(\mathbf{N}, 0.1) = 0$, $res_{pr}(\mathbf{N}, 0.2) = 0$, $res_{pr}(\mathbf{N}, 0.5) = 2$, $res_{pr}(\mathbf{N}, 0.9) = 3$.

For modified bridge: $res_{pr}(\mathbf{N}, 0.1) = 0$, $res_{pr}(\mathbf{N}, 0.2) = 2$, $res_{pr}(\mathbf{N}, 0.5) = 3$, $res_{pr}(\mathbf{N}, 0.9) = 4$.

Comparing $res_{pr}(\mathbf{N}, \beta)$ for different β we can say that modified bridge is more resilient.

Fig. 3.7 Modified bridge

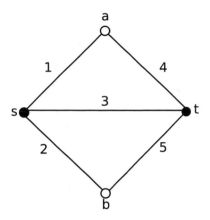

3.7 Static and Dynamic Reliability

Let us have a look at one of the formulas expressing network reliability as a function of element *down* and/or *up* probabilities p_i, q_i. For example, here is the formula for bridge reliability

$$R_0 = 2p^2 + 2p^3 - 5p^4 + 2p^5.$$

In the definition of p or p_i, the **time** factor is **not present**. The reader can imagine that we have a lottery which determines with probability p_i that the state of element i is *up*, and if the element is not *up* then it is *down*. Then we "build" from all elements our network which eventually turns out to be *UP* or *DOWN*. We learned how to compute $P(UP)$, and the formula for R_0 is one of such formulas. What we have obtained is so-called "static" probability to be *UP* which is not connected to any point t on the time axis. If an element became *up* in the lottery, it remains in it "forever". This is why we call it "static".

Now let us introduce **time**. Suppose that each element i has random lifetime τ_i which has continuous distribution $G_i(t) = p(\tau_i \leq t)$. In other words, $P(\tau_i > t) = 1 - G_i(t) = p_i(t)$, which means that with probability $1 - G_i(t)$ element i lifetime **exceeds** t. Therefore at time instant t with probability $G_i(t) = q_i$ the element is *down*. (We neglect the zero probability that $\tau = t$). So, for the instant t, the situation remains the same as it was in static situation. But now we know what will be *after* time t. For example, at the instant $t + \delta$, the element will remain *up* with probability $1 - G(t + \delta)$. And before the instant t the element is *down* with probability $G_i(t)$. We usually assume that element lifetimes are independent random variables.

Summing up, if we have an expression for system *UP* as a function of p_i and q_i, replace p_i by $1 - G_i(t)$ and q_i by $G_i(t)$ and we will obtain the probability $P(T > t)$ that network lifetime T exceeds t. In this way we obtain the picture of network probabilistic behaviour in time, or in short—the network dynamic reliability.

Consider now an example of calculating dynamic reliability.

Assume that all edge lifetimes in bridge network (Fig. 3.2) are independent and identically distributed random variables with $p_i(t) = p(t)$. Then

$$R(t) = P(T > t) = 2p^2(t) + 2p^3(t) - 5p^4(t) + 2p^5(t).$$

Suppose that the lifetime τ of each edge has the same exponential distribution: $\tau \sim exp(\lambda)$. Then

$$R(t) = P(T > t) = 2e^{-2\lambda t} + 2e^{-3\lambda t} - 5e^{-4\lambda t} + 2e^{-5\lambda t}$$

Remind that memoryless property of exponential random variable (see 1.2.9) means that if $V \sim exp(\lambda)$, then $P(V \le t + x | V > t) = P(V < x)$, or equivalent to this property: $P(V > t + x | V > t) = P(V > x)$.

It is worth noting that the exponential distribution is the only distribution having the memoryless property. However network lifetime usually is not distributed exponentially, and therefore this property does not hold for bridge network.

Let us demonstrate this by the following example. Consider again the bridge network and compute the probability that at the instant $t = 3$ the network is UP given that at the instant $t = 1$ it was UP. Assume that $\lambda = 1$. Then

$$P(T > 3 | T > 1) = \frac{P(T > 3)}{P(T > 1)} = \frac{R(3)}{R(1)}$$

$$= \frac{2e^{-2\cdot3} + 2e^{-3\cdot3} - 5e^{-4\cdot3} + 2e^{-5\cdot3}}{2e^{-2\cdot1} + 2e^{-3\cdot1} - 5e^{-4\cdot1} + 2e^{-5\cdot1}} \approx 0.017.$$

Calculating $R(2)$ we get 0.04, i.e. the fact that at the instant $t = 1$ the network was in UP affects its lifetime.

More information on the subject of this chapter can be found in the literature, e.g. in [5, 6].

References

1. Gertsbakh, I. (2000). *Reliability theory with applications to preventive maintenance*. Berlin: Springer.
2. Gertsbakh, I., & Shpungin, Y. (2009). *Models of network reliability: analysis, combinatorics and Monte Carlo*. Boca Raton: CRC Press.
3. Brandes, T., & Erlebach, U. (2005). Network analysis: methodological foundation. Berlin: Springer.
4. Gertsbakh, I., & Shpungin, Y. (2011). *Network reliability and resilience*. Berlin: Springer.
5. Barlow, R. E., & Proschan, F. (1981). *Statistical theory of reliability and life testing*. To Begin With; Reprint Edition.
6. Gertsbakh, I., & Shpungin, Y. (2012). Stochastic models of network survivability. *Quality Technology of Quantitative Management*, 9(1), 45–58.

Chapter 4
Elementary Methods for Reliability Evaluation

Abstract This chapter describes elementary approaches to network reliability evaluation. The first method is based on making a list of all 2^n binary network states. This method is very good but is applicable only to rather small networks with $n = 4 - 6$ elements. For $n > 6$ we need a computer to make the list of all states. Next in line is so-called Crude Monte Carlo (CMC). It consists of performing a series of M lottery-type random experiments of modelling element states and analysing the network state *(UP/DOWN)* as the result of this experiment. If we observe M_1 times the network in *DOWN* state, M_1/M is the CMC estimate of network *DOWN* probability. CMC has, however, an algorithmic complication in identifying the random lottery result as the network *DOWN* or *UP* state. We describe in Sect. 4.5 a technique called DSS (disjoint set structures) for simple and fast identification of network state. During the first reading of this chapter, Sect. 4.5 can be omitted and studied only when the reader starts writing computer codes. This chapter contains also estimation of rel.err. of the CMC, and demonstrates how to analyse network reliability in time (dynamic reliability).

Keywords CMC · State enumeration · DSS · rel.err. · Dynamic reliability

4.1 Exact Method for Computing Network Reliability

In this section, we present the **all state enumeration** method, for **exact** calculation of network reliability. As noted in Sect. 3.1, this method works good for very small networks having 4–6 elements. If the number of elements is $n > 6$ you need a computer to make the list of all 2^n states and for finding the network *UP* states (Table 4.1). For better explanation, we provide here a copy of Table 3.1.

Recall that each binary vector (x_1, x_2, x_3, x_4) in the table represents the state of the network. The sequence of these vectors is constructed in such a way that each subsequent vector is obtained from the previous one by using the binary sum of it and vector $(0, 0, 0, 1)$. Thus, there is no need to keep in memory all the states, but simply to move from each state vector to the next. Directly from the table it is also clear

© The Author(s), under exclusive license to Springer Nature Singapore Pte Ltd. 2020   35
I. Gertsbakh and Y. Shpungin, *Network Reliability*,
SpringerBriefs in Electrical and Computer Engineering,
https://doi.org/10.1007/978-981-15-1458-6_4

Table 4.1 State vectors of network on Fig. 3.1

y	x_1	x_2	x_3	x_4	$\phi(\mathbf{y})$	$P(\mathbf{y})$
$\mathbf{y}(1)$	0	0	0	0	0	$q_1 q_2 q_3 q_4$
$\mathbf{y}(2)$	0	0	0	1	0	$q_1 q_2 q_3 p_4$
$\mathbf{y}(3)$	0	0	1	0	0	$q_1 q_2 p_3 q_4$
$\mathbf{y}(4)$	0	0	1	1	1	$q_1 p_2 p_3 p_4$
$\mathbf{y}(5)$	0	1	0	0	0	$q_1 p_2 q_3 q_4$
$\mathbf{y}(6)$	0	1	0	1	0	$q_1 p_2 q_3 4_4$
$\mathbf{y}(7)$	0	1	1	0	0	$q_1 p_2 p_3 q_4$
$\mathbf{y}(8)$	0	1	1	1	1	$q_1 p_2 p_3 p_4$
$\mathbf{y}(9)$	1	0	0	0	0	$p_1 q_2 p_3 q_4$
$\mathbf{y}(10)$	1	0	0	1	0	$p_1 p_2 p_3 p_4$
$\mathbf{y}(11)$	1	0	1	0	0	$q_1 p_2 p_3 q_4$
$\mathbf{y}(12)$	1	0	1	1	1	$p_1 q_2 p_3 p_4$
$\mathbf{y}(13)$	1	1	0	0	0	$q_1 p_2 p_3 q_4$
$\mathbf{y}(14)$	1	1	0	1	1	$p_1 q_2 p_3 p_4$
$\mathbf{y}(15)$	1	1	1	0	0	$p_1 p_2 p_3 q_4$
$\mathbf{y}(16)$	1	1	1	1	1	$p_1 p_2 p_3 p_4$

how the probability of each state is calculated. The enumeration method scheme is presented in the following Algorithm 4.1.

Assume that the network has n unreliable elements (nodes or edges).

Let $m = 2^n$. Denote by $Y(i)$ the vector state i and by $P(i)$ the **Sum** of UP state probabilities from the $Y(2)$ to $Y(i)$. ($Y(1)$ is the zero vector.) Denote by D_i the set of all *down* elements in $Y(i)$ and by U_i the set of all *up* elements in $Y(i)$.

Algorithm 4.1 Netrel-Enumeration.
   1. Set $Y(1) = (0, \ldots, 0)$ and $P(1) = 0$ //Initialization
   2. For $i = 2$ to m // Cycle 3.-4.
      3. $Y(i) = Y(i-1) + (0, \ldots, 1)$
      4. If $Y(i)$ is UP **Then**

$$P(i) = P(i-1) + \prod_{j \in U(i)} p_j \cdot \prod_{k \in D(i)} q_k$$

If for all elements $p_i = p$, then the enumeration method allows obtaining an **expression** for reliability. Algorithm 4.2 shows the corresponding scheme.

Define an array $A[1, \ldots, n]$ so that $A[i]$ contains a number of UP states with i elements *up* and $n - i$ elements *down*. Denote by One_i number of ones in $Y(i)$.

Algorithm 4.2 Reliability Expression.
   1. Set $Y(1) = (0, \ldots, 0)$. $A[] = (0, \ldots, 1)$.   //Initialization
   2. For $i = 2$ to $m - 1$

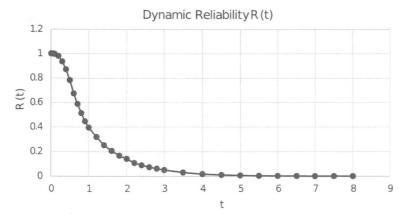

Fig. 4.1 Dynamic reliability function for dodecahedron network

3. $Y(i) = Y(i - 1) + (0, \dots, 1)$
4. If $Y(i)$ is UP **Then** $A[One_i] = A[One_i] + 1$

Using the obtained values in the array A, we can get the formula for network reliability:

$$R = \sum_{i=2}^{n} A[i] p^i q^{n-i} \tag{4.1.1}$$

For example, for the network in Fig. 3.1, we get $A[] = (0, 1, 3, 1)$, and we can write the network reliability in the form $R = p^2 q^2 + p^3 q + p^4$.

Remark 4.1 It will not be superfluous to emphasize once again that Algorithms 4.1 and 4.2 do not require storing the network states table in the computer's memory.

4.2 Crude Monte Carlo (CMC)

Wikipedia gives the following definition of Monte Carlo methods:

"Monte Carlo methods, or Monte Carlo experiments, are a broad class of computational algorithms that rely on repeated random sampling to obtain numerical results. The underlying concept is to use randomness to solve problems that might be deterministic in principle. They are often used in physical and mathematical problems and are most useful when it is difficult or impossible to use other approaches".

In our case, the calculation of network reliability manually or on a computer using a deterministic algorithm, may become difficult or impossible due to the size of the network.

In short, obtaining a network reliability estimate using Monte Carlo can be described as follows:

(a) simulate network state (UP or $DOWN$) M times;
(b) take for the approximate value of reliability the number of UP states divided by the number M.

The corresponding Monte Carlo simulation scheme is presented in the following Algorithm 4.3.

Algorithm 4.3 Netrel-CMC.
  1. Set $Y = 0$
  2. For each element i, **Simulate** its state with probability p_i
  3. Check the network state in accordance with given criterion
  4. If the network state is UP **Then** $Y := Y + 1$
  5. Repeat steps 2,3,4 M times.
  6. Estimate $\hat{R}$ as $\hat{R} = \frac{Y}{M}$

Two questions arise in the presented algorithm: how to simulate element state and how to check the network state. The answer to the first question is simple:

1. **Call function rand()** or any similar function which generates pseudo-random numbers in $(0, 1)$. (Such function exists in any programming language and also in Excel)
2. Assume that the function returned number $a \in (0, 1)$. Then **If** $a \leq p_i$ **Then** element i is *up*, **Otherwise** i is *down*.

As for the second question—checking the state of the network, the situation is more complicated, and we will consider it in Sect. 4.5. Consider now an elementary example of applying Algorithm 4.1. Table 4.2 presents 5 iterations of Algorithm 4.1 for bridge network with non reliable edges (Fig. 3.2).

In columns a_i, for each edge i are presented the values of random variable obtained using **rand()**. Columns x_i contain indicators of the edge i state. Column $\Phi(X)$ contains the structure function values. Consider, for example, the first row of the table. We see that the values of a_1, a_2, a_3 are less than 0.7, and the values a_4, a_5 are greater than 0.7. Hence $x_1 = x_2 = x_3 = 1$, $x_4 = x_5 = 0$, and the network is $DOWN$. From the results of the table we estimate $\hat{R} = \frac{Y}{M} = \frac{3}{5}$.

Table 4.2 Network states simulation ($p_i = p = 0.7$)

k_1	a_1	x_1	a_2	x_2	a_3	x_3	a_4	x_4	a_5	x_5	$\Phi(X)$
1	0.49	1	0.05	1	0.19	1	0.82	0	0.75	0	0
2	0.54	1	0.56	1	0.95	0	0.09	1	0.19	1	1
3	0.58	1	0.004	1	0.17	1	0.43	1	0.69	1	1
4	0.76	0	0.87	0	0.11	1	0.98	0	0.17	1	0
5	0.30	1	0.84	0	0.97	0	0.57	1	0.19	1	1

4.3 Relative Error of CMC

Imagine that we want to estimate the number α by repeated and independent observations of random variable Y which is equal 1 with probability α and 0 with probability $1 - \alpha$. Note that the average value of Y is

$$E[Y] = 1 \cdot \alpha + 0 \cdot (1 - \alpha) = \alpha.$$

Natural unbiased estimator of α is the random variable

$$Z = \frac{Y_1 + Y_2 + \cdots + Y_M}{M}$$

Then

$$E[Z] = \frac{E[Y_1 + \cdots + Y_M]}{M} = \alpha.$$

Let us check the variance and standard deviation of Z:

$$Var[Z] = \frac{M \cdot Var[Y]}{M^2} = \frac{\alpha(1 - \alpha)}{M} \tag{4.3.1}$$

The quality of an estimator is usually measured by the so-called **relative error** (*rel.err.*) which is the standard deviation (the square root of variance) divided by mean. In our case

$$rel.err. = \frac{\sqrt{1 - \alpha}}{\sqrt{M \cdot \alpha}} \tag{4.3.2}$$

A problem appears if α is small. The last equation shows that when α tends to zero, M should tend to infinity fast enough to keep the *rel.err.* small. Suppose $\alpha = 0.001$, and we want to have *rel.err.* $= 0.01-$ one percent. Then after little algebra we obtain that $M = 10^7$-ten million!

It is clear that all of the above is directly related to network reliability evaluation by Monte Carlo. In many cases, the studied networks are very reliable, that is, the values of R are close to 1, and, accordingly, $Q = 1 - R$ is close to 0. In this case, the relative error takes the following form:

$$r.e. = \frac{\sqrt{(1 - Q)}}{\sqrt{M \cdot Q}} \tag{4.3.3}$$

In many cases, we do not have a good preliminary estimate of the value of Q. Therefore for practical purposes, we try to estimate the relative error based on a sample of size K. Each value of this sample represents a network unreliability calculated using M Monte Carlo experiments.

Table 4.3 r.e. with respect to reliability $\hat{R}(q)$ and unreliability $\hat{Q}(q)$ for different values of q for the dodecahedron network

q	$\hat{R}(q)$	r.e. $(\hat{R}(q))$ in %	$\hat{Q}(q)$	r.e. $(\hat{Q}(q))$ in %
0.5	0.59060	0.77	0.40940	1.1
0.4	0.75667	0.79	0.24333	2.5
0.3	0.89877	0.38	0.10123	3.4
0.2	0.97415	0.17	0.02585	6.4
0.1	0.99771	0.075	0.00229	32.8
0.05	0.99975	0.014	0.00025	57.2
0.04	0.99985	0.02	0.00015	134
0.03	0.99997	0.0048	0.00003	161
0.02	0.99999	0.0032	0.00001	316
0.01	1	0	0	∞

Suppose the following K network unreliability values are obtained: $Q_1, Q_2, \ldots, Q_K$. Denote

$$\overline{Q} = \frac{\sum_{i=1}^{K} Q_i}{K}, \quad S^2 = \frac{1}{K-1} \sum_{i=1}^{K} (Q_i - \overline{Q})^2 \tag{4.3.4}$$

Then r.e. is calculated as

$$r.e. = \frac{S}{\overline{Q}} \tag{4.3.5}$$

Consider, for example, the network on Fig. 3.6. Below are 10 network unreliability values calculated by CMC, with $M = 10,000$, $q = 0.1$.
0.00178, 0.00239, 0.00297, 0.00176, 0.00204,
0.00149, 0.00160, 0.00335, 0.00347, 0.00172.
Substituting these values in (4.3.4) we get $\overline{Q} = 0.00226$, $S = 0.00075$. Finally, by (4.3.5) $r.e. = 0.332$, and in percent $r.e. = 33.2\%$.

Table 4.3 shows the relative error values with respect to network reliability and unreliability. For the corresponding calculations, the values of $M = 10,000$, $K = 10$ were chosen.

Remark 4.2 In the case that we are not satisfied with the relative error and want to reduce it by k times, it is enough to increase M by k^2 times. This recommendation follows from formula (4.3.3).

4.4 CMC for Network Dynamic Reliability Evaluation

As explained in Sect. 3.7, there is a simple relationship between the expressions of static and dynamic reliability. Namely, if we have an expression for network UP as a function of p_i and q_i, replace p_i by element i dynamic reliability $r_i(t) = P(\tau_i > t)$, and q_i by $1 - r_i(t)$, and we will obtain the network dynamic reliability.

Thus, taking Algorithm 4.3 as a basis, we arrive at the following Monte Carlo scheme for dynamic reliability.

Algorithm 4.4 Net-dynamic-rel-CMC.
   1. Determine the time instants $t_1, t_2, \ldots, t_l$ at which we want to get the values of $R(t)$.
   2. For each time instant $t_j, j \in [1, l]$, **Do** steps 3-9.
   3. For each element i calculate $p_i = r_i(t_j)$. So we have **static** probabilities for all elements.
   4. Set $Y = 0$.
   5. For each element i, **Simulate** its state with probability p_i.
   6. Check the network state in accordance with given criterion.
   7. If the network state is UP **Then** $Y := Y + 1$.
   8. Repeat steps 5,6,7 M times.
   9. Estimate $\hat{R}(t_j)$ as $\hat{R}(t_j) = \frac{Y}{M}$.

Table 4.4 presents the simulation results for the dodecahedron network with M = 10,000. For simplicity, we assumed that for all elements their lifetimes are exponentially distributed with $\lambda = 1$: $r_i(t) = exp(-t)$. The corresponding graph is shown in Fig. 4.1.

Table 4.4 Dynamic reliability $R(t)$ obtained by CMC for the dodecahedron network

j	t_j	p_j	$\hat{R}(t_j)$
1	0.01	0.99	1
2	0.05	0.95	0.9994
3	0.1	0.90	0.9972
4	0.5	0.0.61	0.7812
5	0.8	0.45	0.5128
6	1	0.37	0.3947
7	1.6	20	0.2053
8	2	0.14	0.1410
9	3	0.05	0.0500
10	5	0.007	0.007
11	8	0.0003	0.0003

4.5 Checking Network State

In this section we give some recommendations how check the network state. Clearly, checking depends on the *UP–DOWN* criterion. There are many algorithms that serve this purpose. We will describe a simple method that fits most of the criteria we considered. It is based on so-called *Disjoint Set Structures* (DSS), see [1, 2]. Before describing the algorithm, we will explain its idea using the following example.

In the network on Fig. 4.2a the edges are unreliable, and the criterion is three terminals connectivity. Assume that the network is in the state shown in Fig. 4.2b. The checking process begins with the zero state shown in Fig. 4.2c. Assign a sequence number to each node, and say that the node belongs to the set with this number. For example:

$$\text{nodes: a b c d e f}$$
$$\text{sets: 1 2 3 4 5 6}$$

Now we will add *up*—edges from a given state in arbitrary order, for example: 1, 3, 4, 8, 2, 5. After adding edge 1, a non-trivial component of size 2 has been formed (Fig. 4.2d). Let **merge** two sets containing nodes *a* and *b* into one set with number 1 or 2 (it does not matter):

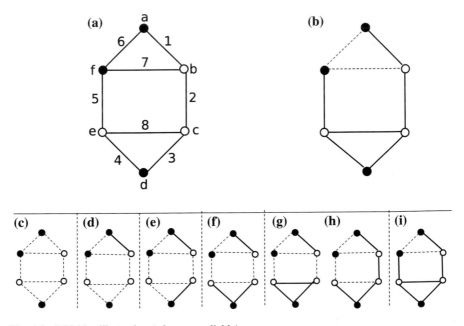

Fig. 4.2 DSS idea illustration (edges nonreliable)

$$\text{nodes: } (a, b) \ c \ d \ e \ f$$
$$\text{sets: } \quad 1 \quad 3 \ 4 \ 5 \ 6$$

Continue in the same way, checking at each stage whether all terminals are in the same set. If yes, it means that the state is UP. Otherwise it is $DOWN$. After adding edges 3 and 4, we get the following structure:

$$\text{nodes: } (a, b) \ (c, d, e) \ f$$
$$\text{sets: } \quad 1 \quad \quad 3 \quad \quad 6$$

The next edge is 8. Its nodes are c and e. Both belong to the set 3. This means that this time the new set has not been created, and there is no need to merge the sets. Therefore, disjoint set structure **remains the same** and the state Fig. 4.2g is still $DOWN$. After adding edge 2 we get the $DOWN$ state Fig. 4.2h and the following set structure:

$$\text{nodes: } (a, b, c, d, e) \ f$$
$$\text{sets: } \quad \quad 1 \quad \quad 6$$

And finally, after adding edge 5, all three terminals are found in one set:

$$\text{nodes: } (a, b, c, d, e, f)$$
$$\text{sets: } \quad \quad 1$$

So, the given state is UP.

Remark 4.3 The checking process continues either until all terminals will be in the same set or/and all edges of the given state are added. For example, if in our network were defined two terminals—a and b, then Fig. 4.2h is in UP and the process stops.

Remark 4.4 From the above example it is clear that the DSS algorithm is suitable for other criteria: all nodes connectivity, maximal component, maximal cluster, as well as some other criteria which are not considered by us here. For example, if the criterion were the "size of the maximal cluster should be at least 3", then Fig. 4.2f satisfies this requirement and the process stops.

Consider now the case of unreliable nodes. Network in Fig. 4.3a has two terminals. The criterion is terminal connectivity. Recall that if a node is in the $down$ state, then all edges incident to this node are erased, but the node itself remains intact. Suppose that Fig. 4.3b is the state of the network. As in the previous case, we start the DSS process from the zero state. We will add the nodes, for example, in the following order: c, e, f, b. Then, similarly to the previous example, we obtain the following sets of structures:

The zero state Fig. 4.3c:

$$\text{nodes: } a \ b \ c \ d \ e \ f$$
$$\text{sets: } \ 1 \ 2 \ 3 \ 4 \ 5 \ 6$$

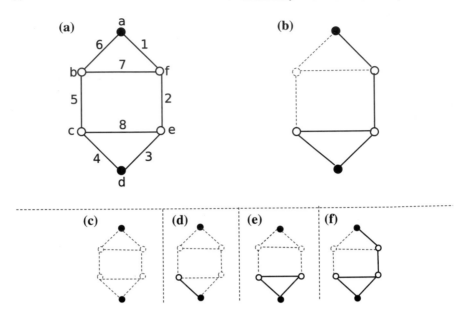

Fig. 4.3 DSS idea illustration (nodes nonreliable)

Adding node c (Fig. 4.3d):

$$\text{nodes: a b (c, d) e f}$$
$$\text{sets: 1 2   3   5 6}$$

Note that no additional edges appeared, in accordance with the definition of the nodes in the *down*. After adding node e we get:

$$\text{nodes: a b (c, d, e) f}$$
$$\text{sets: 1 2   3   6}$$

Finally, adding node f we arrive at the following state Fig. 4.3f.

$$\text{nodes: (a, c, d, e, f) b}$$
$$\text{sets: 1   2}$$

We see that the terminals are in the same set, and therefore the given state is UP.

Remark 4.5 Among criteria mentioned in Chap. 2, were also criteria "minimal path" and "maximal flow". We do not provide here the descriptions of the corresponding algorithms. The first of these is the well-known Dijkstra algorithm (see, for example, [2]). The second will be mentioned in Chap. 8 of this book.

DSS Algorithm

We will need to execute a sequence of operations of two kinds:

1. Given a node, we find out which set contains it, and return the number of this set (component).
2. Given two different labels (numbers of sets), we *merge* the contents of the two corresponding sets, and choose a label for the combined set (say for definiteness, the first one of the two).

These two operations are the main parts of the DSS algorithm. Indeed, when we add the next edge $e = (v, w)$ we check to which components belong its incident nodes v and w. In cases they belong to the same component, we do not need merging. Note that we present here the simplest versions of these operations. Much more efficient implementations can be found, for example, in [2].

Define now an array $Set[1, \ldots, n]$ for each node $i, i = 1, \ldots, n$. $Set[i]$ means the label of the set containing node i.

Function *find* (i)
  Return $Set[i]$

The following function merges the sets labelled a and b.

Function *merge*(a, b)
  For $k = 1$ to n
    If $Set[k] = b$ **Then** $Set[k] = a$
  Return a   //a-the first of two labels

Now we are ready to present the pseudo-code for the DSS algorithm, for both cases of connectivity—the k-terminal connectivity and the overall connectivity. We present two variants of the pseudo-code: the first for unreliable edges, and the second for unreliable nodes. Suppose that the network has n nodes and m edges.

The pseudo-codes use the following arrays.

a1.—for the case of unreliable edges. $EdgeState[1, \ldots, s]$—the array of the given state edges in *up*.

a2.—for the case of unreliable nodes. $NodeState[1, \ldots, l]$—the array of the given state nodes in *up*, **including terminals**.

The remaining arrays are used in both variants.

b. $Fnode[1, \ldots, m]$—the array of all edge first nodes.

c. $Snode[1, \ldots, m]$—the array of all second edge nodes.

So, $EdgeState[i]$ has two nodes: $Fnode[i]$ and $Snode[i]$. We suppose that $Fnode[i] < Snode[i]$.

d. $Set[1, \ldots, n]$—the array of sets for nodes, so that the node i belongs to the set $Set[i]$.

e. $T[n_1, \ldots, n_k]$—the array of terminal numbers.

In particular, this array can contain all the vertices of the network.

f. $Tset[1, \ldots, n]$; $Tset[i]$ gives the number of terminals belonging to the set i.

The following algorithm checks the state of a network with unreliable edges.

Algorithm 4.5 Check-edges-DSS.
// Edges unreliable.
    1. For $i = 1$ to n //Initialization
       $Set[i] = i$
       $Tset[i] = 0$
    2. For $i = 1$ to k
       $j = T[i]$
       $Tset[j] = 1$
    3. For $i = 1$ to s // Check all edges of the given state
       $j = EdgeState[i]$
       $u = Fnode[j]$
       $v = Snode[j]$
       $uset = find(u)$
       $vset = find(v)$
       If ($uset \neq vset$) **Then**   //Check whether these sets are different
           $r = merge(uset, vset)$
           $Tset[r] = Tset[uset] + Tset[vset]$ //Number of terminals
// after merging
       If $Tset[r] = k$ **Then Return**("UP")   //End of 3
    Return ("$DOWN$")

The following algorithm checks the state of a network with unreliable nodes.

Algorithm 4.6 Check-nodes-DSS.
// Nodes unreliable. Suppose that NodeState contains also other nodes
// besides terminals.
    1. For $i = 1$ to n //Initialization
       $Set[i] = i$
       $Tset[i] = 0$
    2. For $i = 1$ to k
       $j = T[i]$
       $Tset[j] = 1$
    3. For $i = 1$ to $l - 1$
       $x = NodeState[i]$
           4. For $j = i + 1$ to l
          $y = NodeState[j]$
          $a = min(x, y); b = max(x, y)$
       //For all pairs of nodes in up, check whether they
       // are connected by any edge.
           If there exists such z that $a \in Fnode[z]$ and $b \in Snode[z]$
              Then $u = Find(a); v = Find(b)$
           If $u \neq v$ **Then**
              $r = Merge(u, v); Tset[r] = Tset[u] + Tset[v]$
              If $Tset[r] = k$ **Then Return**(UP)

       End   // End of **4**
End   // End of **3**
Return($DOWN$)

As previously noted, the DSS algorithm also fits many other criteria defined for components and terminals. So, for example, for criteria the "maximum component" or "maximum cluster", you need to add a variable *sizeOfComp* and make minimal simple changes.

References

1. Gertsbakh, I., & Shpungin, Y. (2009). *Models of Network Reliability: Analysis Combinatorics and Monte Carlo*. Boca Raton: CRC Press.
2. Cormen, T., Leiserson, C., Rivest, R., & Stein, C. (2010) *Introduction to Algorithms*. (3rd ed.). Cambridge: The MIT Press.

Chapter 5
Element Importance

Abstract Birnbaum Importance Measure (BIM) of element j is defined as $\partial R/\partial p_j$, where R is system reliability. BIM_j measures the influence of element j reliability p_j on system reliability R. Element importance analysis plays important role in system reliability design. We consider in this chapter several methods of calculating BIM_j based on Monte Carlo, direct enumeration and structure analysis for some simple structures like series and parallel systems. We demonstrate also the impact of most important element improvement on system reliability.

Keywords BIM · Series/parallel systems · Enumeration · Monte Carlo · Optimal design

5.1 Birnbaum Importance Measure (BIM)

The so-called element *importance* analysis plays a crucial role in network reliability design. This analysis is based on computing element importance measures first suggested by Z. Birnbaum in his work [1].

Suppose the network consists of n independent elements, and element i has reliability p_i, $i = 1, 2, \ldots, n$. Let network reliability be

$$R = \Psi(p_1, p_2, \ldots, p_n). \tag{5.1.1}$$

Importance of element j has been originally suggested in [1] as

$$BIM_j = \frac{\partial \Psi(p_1, p_2, \ldots, p_n)}{\partial p_j} \tag{5.1.2}$$

Abbreviation BIM_j is *Birnbaum importance measure of element j*. Using the pivotal decomposition, see [2, 3] and our Chap. 3, we represent $R = \Psi(p_1, p_2, \ldots, p_n)$ as follows:

$$R = p_j \Psi(p_1, p_2, \ldots, 1_j, \ldots, p_n) + (1 - p_j)\Psi(p_1, p_2, \ldots, 0_j, \ldots, p_n) \tag{5.1.3}$$

© The Author(s), under exclusive license to Springer Nature Singapore Pte Ltd. 2020   49
I. Gertsbakh and Y. Shpungin, *Network Reliability*,
SpringerBriefs in Electrical and Computer Engineering,
https://doi.org/10.1007/978-981-15-1458-6_5

From here it follows that

$$BIM_j = \Psi(p_1, p_2, \ldots, 1_j, \ldots, p_n) - \Psi(p_1, p_2, \ldots, 0_j, \ldots, p_n). \qquad (5.1.4)$$

So, by (5.1.4) we see that BIM has a transparent probabilistic meaning: it is the *gain* in network reliability received from replacing a *down* element j by an absolutely reliable one. It should also be noted that the following approximate equality follows from (5.1.2):

$$\Delta R \approx BIM_j \delta p_j, \qquad (5.1.5)$$

where ΔR is network reliability increment resulted from element j reliability increment by δp_j.

BIM is very essential for solving various problems of optimal network design. Some examples of such design will be presented in Chap. 8.

5.2 Direct Computation of BIM

For direct computation of element's BIM we need a formula for network reliability R as a function of element reliabilities p_i.

Series and Parallel Systems

Consider first a series system with n elements having *up* probabilities $p_1 > p_2 > \cdots > p_n$. Then

$$R_{ser} = \Pi_{j=1}^n p_j.$$

Then $\partial R_{ser}/\partial p_j = \Pi_{j=1}^n p_j/p_j$. From this formula follows that the largest BIM has the element having the smallest reliability. The **less reliable** element is the most important in the series system!

Suppose we have a parallel system of elements with reliabilities $p_1 > p_2 > \cdots > p_n$. The reliability of this system is

$$R_{par} = 1 - \Pi_{j=1}^n (1 - p_j).$$

Then

$$\partial R_{par}/\partial p_j = \Pi_{j=1}^n (1 - p_j)/(1 - p_j)$$

From this formula follows that the **maximal reliable** element has the maximal BIM_j.

Small Networks

Consider now the network in Fig. 5.1, and calculate BIM_j for all $j = 1, \ldots, 5$.

The set of all minimal paths is (3), (1, 4), (2, 5), and using (3.3.1) we get

$$R = p_3 + p_1 p_4 + p_2 p_5 - p_1 p_3 p_4 - p_2 p_3 p_5 - p_1 p_2 p_4 p_5 + p_1 p_2 p_3 p_4 p_5$$

From the expression of R we get the following $BIM's$ for all edges:

$$BIM_1 = \frac{\partial R}{\partial p_1} = p_4 - p_3 p_4 - p_2 p_4 p_5 + p_2 p_3 p_4 p_5$$

$$BIM_2 = \frac{\partial R}{\partial p_2} = p_5 - p_3 p_5 - p_1 p_4 p_5 + p_1 p_3 p_4 p_5$$

$$BIM_3 = \frac{\partial R}{\partial p_3} = 1 - p_1 p_4 - p_2 p_5 + p_1 p_2 p_4 p_5$$

$$BIM_4 = \frac{\partial R}{\partial p_4} = p_1 - p_1 p_3 - p_1 p_2 p_5 + p_1 p_2 p_3 p_5$$

$$BIM_5 = \frac{\partial R}{\partial p_5} = p_2 - p_2 p_3 - p_1 p_2 p_4 + p_1 p_2 p_3 p_4$$

From the above formulas it is clear that the values of p_j affect the values of $BIM's$. But besides this there is another factor—topological. To explain this, assume that for all edges $p_j = p$. Then we get the following formulas for $BIM's$:

$$BIM_1 = BIM_2 = BIM_4 = BIM_5 = p - p^2 - p^3 + p^4$$
$$BIM_3 = 1 - 2p^2 + p^4$$

Comparing the obtained expressions we see that $BIM_3 > BIM_j$ for all $j \neq 3$ and all p. So, the edge 3 is the most important. Notice, that considering Fig. 5.1, this conclusion is also supported by our intuition.

Consider now the following problem. Suppose that for all edges $p_j = 0.8$, and we have the ability to replace one edge with a more reliable one, say with $p = 0.9$. The question is what edge to replace to get the maximal increase of network reliability.

Fig. 5.1 Modified bridge

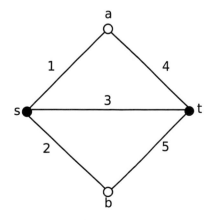

The answer is obvious: edge 3. We have $BIM_3 = 1 - 2 \cdot 0.8^2 + 0.8^4 = 0.1296$. In this case, the increase in reliability will be, by 5.1.5, $\Delta R \approx 0.1296 \cdot 0.1 = 0.01296$.

Consider now another case of element reliability: $p_3 = 0.5$, $p_1 = p_2 = p_4 = p_5 = 0.8$. In this case, we get: $BIM_3 = 0.1296$, $BIM_j = 0.144$, $j = 1, 2, 4, 5$. Now edge 3 does not have maximal BIM. And yet the replacement of the edge 3 gives the maximal increment: $\Delta R = 0.1296 \cdot (0.9 - 0.5) \approx 0.052$ for edge 3 and $\Delta R = 0.144 \cdot (0.9 - 0.8) \approx 0.014$ for edge j, $j = 1, 2, 4, 5$. The case when the topological factor prevails is not mandatory, but it occurs quite often. More details about the properties and application of BIM will be given in Chaps. 6 and 8.

Computing BIM by Enumeration

As noted in Chap. 3, using the enumeration method can be very useful for computing reliability for networks of moderate size. For example, this method is applicable to a network with 20 unreliable nodes and 50 reliable edges. Recall that this method gives accurate results.

Table 5.1 is a part of Table 3.1 (describing all network states) and represents network UP states, with the corresponding probabilities. We have by definition:

$$BIM_j = \frac{\partial P(Y(4))}{\partial p_j} + \frac{\partial P(Y(8))}{\partial p_j} + \frac{\partial P(Y(12))}{\partial p_j} + \frac{\partial P(Y(14))}{\partial p_j} + \frac{\partial P(Y(16))}{\partial p_j}$$

Chapter 4 gave pseudo-codes for calculating reliability using enumeration, as well as for deriving an expression for reliability, in the case of equiprobable elements.

The enumeration method scheme for BIM calculation is presented in the following Algorithm 5.1.

Assume that the network has n unreliable elements (nodes or edges). Denote by $Y(i)$ the vector state number i. Let $BIM[1, \ldots, n]$ be an array so that $BIM[j]$ will contain BIM_j.

Denote by D_i the set of all *down* elements in $Y(i)$ and by U_i the set of all *up* elements in $Y(i)$.

Table 5.1 UP state vectors of network on Fig. 3.1

y	x_1	x_2	x_3	x_4	$P(y)$
y(4)	0	0	1	1	$q_1 q_2 p_3 p_4$
y(8)	0	1	1	1	$q_1 p_2 p_3 p_4$
y(12)	1	0	1	1	$p_1 q_2 p_3 p_4$
y(14)	1	1	0	1	$p_1 p_2 q_3 p_4$
y(16)	1	1	1	1	$p_1 p_2 p_3 p_4$

Table 5.2 Application of Algorithm 5.1 to network on Fig. 3.1

$Y(i)/BIM[k]$	$BIM[1]$	$BIM[2]$	$BIM[3]$	$BIM[4]$
$Y(4)$	$-q_2 p_3 p_4$	$-q_1 p_3 p_4$	$q_1 q_2 p_4$	$q_1 q_2 p_3$
	$+$	$+$	$+$	$+$
$Y(8)$	$-p_2 p_3 p_4$	$q_1 p_3 p_4$	$q_1 p_2 p_4$	$q_1 p_2 p_3$
	$+$	$+$	$+$	$+$
$Y(12)$	$q_2 p_3 p_4$	$-p_1 p_3 p_4$	$p_1 q_2 p_4$	$p_1 q_2 p_3$
	$+$	$+$	$+$	$+$
$Y(14)$	$q_2 p_3 p_4$	$-p_1 p_3 p_4$	$p_1 q_2 p_4$	$p_1 p_2 q_3$
	$+$	$+$	$+$	$+$
$Y(16)$	$p_2 p_3 p_4$	$p_1 p_3 p_4$	$p_1 p_2 p_4$	$p_1 p_2 p_3$

Algorithm 5.1 BIM-Enumeration.

   1. Set $Y(1) = (0, \ldots, 0)$ and $BIM[] = (0, \ldots, 0)$ //Init
   2. For $i = 2$ to n // Cycle 3.-5.
      3. $Y(i) = Y(i - 1) + (0, \ldots, 1)$.
      4. Find D_i and U_i.
      5. If $Y(i)$ is UP **Then**
         For $k = 1$ to n
            If $k \in U_i$ **Then** $x = 1$ **Else** $x = -1$

$$BIM[k] = BIM[k] + \prod_{j \in U_i, j \neq k} p_j \cdot \prod_{s \in D_i, s \neq k} q_s \cdot x$$

Table 5.2 shows the application of the algorithm to the network on Fig. 3.1, in accordance with changes of i.

If for all elements $p_i = p$, then the enumeration method allows to obtain an **expression** for all $BIM's$. Algorithm 5.2 shows the corresponding scheme.

Define a two-dimensional array $A[1, \ldots, n][0, \ldots, n - 1]$ so that $A[k, j]$ will contain the number of terms with j elements up, in the expression for the partial derivative $\frac{\partial R}{\partial p_k}$. Denote by $Ones_i$ number of ones in $Y(i)$.

Algorithm 5.2 BIM Expression.

   1. Set $Y(1) = (0, \ldots, 0)$, $A[][] = (0, \ldots, 0)(0, \ldots, 0)$ //Init
   2. For $i = 2$ to n
      3. $Y(i) = Y(i - 1) + (0, \ldots, 1)$
      4. Find D_i and U_i.
      5. If $Y(i)$ is UP **Then**
         For $k = 1$ to n
            If $k \in U_i$ **Then** $A[k, Ones_i - 1] = A[k, Ones_i - 1] + 1$
               Else $A[Ones_i] = A[Ones_i] - 1$

Table 5.3 Example of getting an expression for BIM_4

Y/A	A[4, 0]	A[4, 1]	A[4, 2]	A[4, 3]
y(4)		+1		
y(8)			+1	
y(12)			+1	
y(14)			+1	
y(16)				+1
Sum		1	3	1

Using the obtained values in the array $A[][]$, we can get the expression for BIM:

$$BIM_k = \sum_{i=0}^{n-1} A[k, i] p^i q^{n-1-i} \qquad (5.2.1)$$

In Table 5.3 we see an example of calculations for BIM_4 using Algorithm 5.2. So, we have $BIM_4 = pq^2 + 3p^2q + p^3$.

5.3 Monte Carlo for Computing BIM

Reminder. An explanation of Monte Carlo network reliability evaluation was given in Sect. 4.2.

Consider the dodecahedron shown in Fig. 5.2. Vertices 1, 14 and 19 are terminals. Assume that all edges are reliable, and the reliability of each node is equal to $p = 0.7$.

Reliability calculation by CMC gives $R = 0.82$. Recall that according to formula 5.1.4

$$BIM_j = R(p_1, \ldots, 1_j, \ldots, p_n) - R(p_1, \ldots, 0_j, \ldots, p_n)$$

Table 5.4 presents the importance of all vertices. We see that four nodes—4, 5, 13, 20 are of maximal importance. In the table, these nodes are marked with an asterisk in the row j. This result is also supported by the following "non-trivial" intuitive reasoning:

Each of these nodes has the following two properties:

(1) the node belongs to one of the minimal cuts;

(2) the node is at a distance of 1 from some one of the terminals, and at a distance of 2 from some other terminals.

As previously noted, element importance analysis plays a significant role in network reliability design. Examples of such design are given in Chap. 8, and can be found in [3, 4].

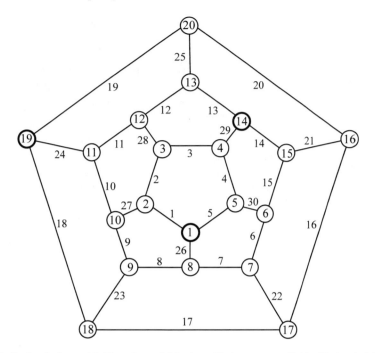

Fig. 5.2 Dodecahedron with 20 nodes and 30 edges. Nodes are not reliable. Nodes 1, 14, 19 are terminals

Table 5.4 Dodecahedron BIM's calculations. Nodes unreliable. Terminals T = (1, 14, 19). $p_j =$ 0.7 for all j

j	2	3	4*	5*	6	7	8	9	
$R(1_j)$	0.86	0.84	0.87	0.87	0.84	0.84	0.86	0.84	
$R(0_j)$	0.72	0.77	0.71	0.71	0.77	0.78	0.73	0.77	
BIM_j	0.14	0.07	0.16	0.16	0.07	0.06	0.13	0.07	
j	10	11	12	13*	15	16	17	18	20*
$R(1_j)$	0.84	0.86	0.84	0.87	0.86	0.84	0.84	0.86	0.87
$R(0_j)$	0.77	0.72	0.77	71	72	0.77	0.78	0.73	0.71
BIM_j	0.07	0.14	0.07	16	0.14	0.07	0.06	0.13	0.16

Here we limit ourselves by the following simple example. Let us assume that we are interested in increasing the reliability of the network, and we have the opportunity to replace 3 nodes with more reliable ones, say with *up* probability $p = 0.9$. It is clear that in order to optimally replace one node, it is necessary to select one of the nodes with maximal BIM. In case of replacing several elements, the best solution is not necessarily the choice of the best elements. However, this heuristic approach is quite good, and we will use it now. So we choose to replace for example the vertices 4, 5, 13 by more reliable with $p = 0.9$. Monte Carlo calculations show that in this

Table 5.5 Dodecahedron BIM's calculations. Nodes unreliable. Terminals T=(1, 14, 19). $p_j =$ 0.7 for $j \in [2, 9]$, $p_j = 0.8$ for the rest of j

j	2*	3	4	5*	6	7	8*	9	
$R(1_j)$	0.94	0.92	0.93	0.94	0.91	0.92	0.94	0.92	
$R(0_j)$	0.80	0.87	0.84	0.81	0.87	0.87	0.81	0.87	
BIM_j	0.14	0.05	0.09	0.13	0.04	0.05	0.13	0.05	
j	10	11	12	13	15	16	17	18	20
$R(1_j)$	0.91	0.92	0.91	0.92	0.92	0.91	0.91	0.92	0.92
$R(0_j)$	0.86	0.83	0.87	82	82	0.87	0.87	0.85	0.83
BIM_j	0.05	0.09	0.04	10	0.10	0.04	0.04	0.07	0.09

case the reliability increases from $R = 0.82$ to $R = 0.91$. It is worth noting that in our example the replacement of not the best nodes leads to a smaller increase in reliability.

Let us now consider the same network, but with the following up probabilities: $p_j = 0.7$ for $j \in [2, 9]$ and $p = 0.8$ for the rest of j.

CMC calculations for this network give $R = 0.90$ and BIM's are presented in Table 5.5. Note that significant changes in node probabilities led to a change in the structure of the best elements. In this network, the most important are nodes 2, 5, 8. However, the topological factor has remained here as well: (2, 5, 8) is the minimal cut.

References

1. Birnbaum, Z. W. (1969). On the importance of different components in multicomponent system. In P. R. Krishnaiah (Ed.), *Multivariate Analysis* (pp. 581–592). New York: Academic.
2. Gertsbakh, I., & Shpungin, Y. (2009). *Models of network reliability: Analysis, combinatorics and Monte Carlo*. Boca Raton: CRC Press.
3. Gertsbakh, I., & Shpungin, Y. (2011). *Network reliability and resilience*. Berlin: Springer.
4. Gertsbakh, I., & Shpungin, Y. (2012). Combinatorial approach to computing importance indices of coherent systems. *Probability in Engineering and Information Sciences, 26*, 117–128.

Chapter 6
Destruction Monte Carlo

Abstract In this chapter we introduce a new and very important network structural invariant—so-called cumulative destruction spectrum (CD-spectrum). This parameter allows to investigate network behaviour when the network is subject to a flow of external "shocks" which randomly destroy network elements (nodes or edges). We illustrate the use of the CD-spectrum by analysing and comparing the behaviour of two small transportation networks having the same number of nodes, edges and terminals but differ in their structure. Section 6.3 is devoted to another structural invariant—the so-called BIM-spectrum. Using this spectrum becomes possible to range network elements (nodes or edges) by their importance. The chapter is concluded by describing in detail a Monte Carlo algorithm for computing the CD-spectrum and the BIM-spectrum.

Keywords Signature · Internal distribution · CD-spectrum · BIM-spectrum · Network comparison

6.1 Individual and Comparative Network Reliability Analysis

All methods for network reliability calculation considered in previous chapters were aimed at the "individual" analysis of a single network. Given node/edge reliability parameters and network UP/DOWN criterion, we were able to carry out reliability analysis for this particular network.

In engineering practice, very often there is a need to consider and compare *several* network-type solutions for solving real-life problems. For example, two different designs of the transportation network have been suggested for the same region, and the engineers have to compare them by their performance parameters. Suppose that both networks have the same number of terminals, nodes and edges and the same UP/DOWN criteria.

We introduce now a new situation that both networks are subject to a random flow of "shock-type" events, e.g. random attacks or earthquakes, which destroy network elements subject to failure. These shocks act randomly, and each shock destroys

© The Author(s), under exclusive license to Springer Nature Singapore Pte Ltd. 2020   57
I. Gertsbakh and Y. Shpungin, *Network Reliability*,
SpringerBriefs in Electrical and Computer Engineering,
https://doi.org/10.1007/978-981-15-1458-6_6

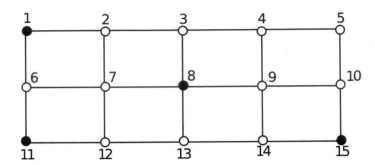

Fig. 6.1 Grid network with 15 nodes and 22 edges. Edges unreliable. Nodes 1, 8, 11, 15 are terminals

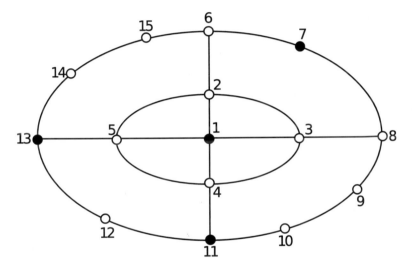

Fig. 6.2 Ring network with 15 nodes and 22 edges. Edges unreliable. Nodes 1, 7, 11, 13 are terminals

with equal probability one of the edges (or nodes) which is "alive" before the shock appearance.

For this shock model, reliability engineers must compare both networks by their behaviour and to find out which design is more resilient (or more reliable). This comparison is complicated by the necessity to consider several UP/DOWN criteria for both networks. To explain better this new situation, let us consider an example.

Let us start with comparing two small railway networks, a grid-type on Fig. 6.1 and a ring-type on Fig. 6.2. Both networks have 15 reliable nodes, 22 unreliable edges and four terminal nodes. We are interested in analysing and comparing both network behaviour in "shock" attack, after random failure of several edges. No one of previously considered methods, e.g. enumeration, is not practical in this new situation.

To solve this new kind of problems, we introduce in this chapter a new formal tool—the so-called *cumulative D-spectrum.*

Definition 6.1 The cumulative D-spectrum (CD-spectrum) is an ordered set of non negative numbers

$$(y_1, y_2, \ldots, y_n), y_1 \le y_2 \le \cdots\cdots \le y_n,$$

where y_i is the probability that the network is in $DOWN$ state after successive random destruction of i of its elements, and n is the number of elements.

Below, in Sect. 6.2, we will describe the CD-spectrum in all detail. Right now we mention, without proof, one **fundamental** property of the CD-spectrum. This property is the relationship between y_k and the number $C(k)$ of cut sets of the network of size k:

$$C(k) = \frac{n!}{k!(n-k)!} y_k, k = 1, 2, \ldots, n \qquad (6.1.1)$$

It is important to stress that the CD-spectrum is a *combinatorial or structural invariant* of the network: it depends only on network structure and does not depend on network element failure mechanism.

In literature, two notions close to CD-spectrum have been introduced in [1, 2]-*signature and Internal Distribution (ID)*. Signature $f_1, f_2, \ldots, f_n$ is connected to the CD-spectrum via the relationship: $f_1 + \cdots + f_k = y_k$ and ID just coincides with the signature.

In the following sections, we compare the networks shown on Figs. 6.1, 6.2 by their behaviour in the process of successive destruction of their elements, for three different criteria:
(a) Terminal connectivity.
(b) All Terminal (all node) connectivity.
(c) Maximal cluster, when the network is UP if there is a cluster of size at least 10 nodes. (For cluster definition see Sect. 2.1.)

Interesting to note that for criterion (a) the ring is better, for criterion (b) a grid is better and the case of criterion (c), the networks have almost identical behaviour.

The corresponding explanations and the results we will see in Sect. 6.4.

6.2 Network Destruction

In this section, we will give a non formal explanation of the network destruction process, the cumulative destruction spectrum, and the calculation of network reliability using this spectrum.

Consider the network shown in Fig. 6.3. It has 6 nodes and 8 edges. The nodes s and t arc terminals, and the remaining nodes are subject to failure.

Fig. 6.3 Network with 6
nodes and 8 edges. Nodes
are unreliable. Nodes s, t are
terminals

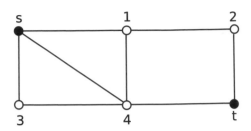

Table 6.1 All node permutations

Column 1	Column 2	Column 3	Column 4
1, 2, 3, 4*	2, 1, 3, 4*	3, 1, 2, 4*	4, 1*, 2, 3
1, 2, 4*, 3	2, 1, 4*, 3	3, 1, 4*, 2	4, 1*, 3, 2
1, 3, 2, 4*	2, 3, 1, 4*	3, 2, 1, 4*	4, 2*, 1, 3
1, 3, 4*, 2	2, 3, 4*, 1	3, 2, 4*, 1	4, 2*, 3, 1
1, 4*, 2, 3	2, 4*, 1, 3	3, 4, 1*, 2	4, 3, 1*, 2
1, 4*, 3, 2	2, 4*, 3, 1	3, 4, 2*, 1	4, 3, 2*, 1

Let us imagine the following random process. Order the nodes which are subject
to failure in random order. The ordinal numbers of these nodes will be a random
permutation i_1, i_2, i_3, i_4.

Let us start an operation called *destruction*. Initially all nodes are in the *up* state,
therefore the network is *UP*. Destroy the nodes from left to right. First, destroy the
node i_1, and check the network state. If it is *DOWN*, stop the process. Otherwise
destroy the second node i_2 and proceed.

Table 6.1 shows all 24 permutations of the nodes. The nodes destruction of which
caused the failure of the network are marked by asterisk. For example, the permutation
$(3, 2, 4, 1)$ (see column 3) shows that the network became *DOWN* on the third step of
the destruction process, after the failure of node 4. In this case, we say that number
3 is the *anchor* of the permutation $(3, 2, 4, 1)$.

In general, for any type of elements and for any *UP/DOWN* criterion, we give
the following definition:

Definition 6.2 Let π be a permutation of unreliable elements: $e_{i_1}, \ldots, e_{i_n}$. Start with
a network with all elements being *up* and "erase" ("destroy") the elements in the
order of their appearance in π, from left to right. Stop at the *first* element e_{i_r}, when
the network becomes *DOWN*. The ordinal number r of this element is called the
anchor of the permutation π and is denoted $r(\pi)$.

Note that anchor value for given π depends only on the *network structure* and its
DOWN definition. It is completely separated from the stochastic mechanism which
governs element failures in a real element failure process.

Table 6.2 Grouping permutations by values of $r(\pi)$

$r(\pi) = 1$	$r(\pi) = 2$	$r(\pi) = 3$	$r(\pi) = 4$
	1, 4, 2, 3	1, 2, 4, 3	1, 2, 3, 4
	1, 4, 3, 2	1, 3, 4, 2	1, 3, 2, 4
	4, 1, 2, 3	2, 1, 4, 3	2, 1, 3, 4
	4, 1, 3, 2	2, 3, 4, 1	2, 3, 1, 4
	2, 4, 1, 3	3, 1, 4, 2	3, 1, 2, 4
	2, 4, 3, 1	3, 2, 4, 1	3, 2, 1, 4
	4, 2, 1, 3	3, 4, 1, 2	
	4, 2, 3, 1	3, 4, 2, 1	
		4, 3, 1, 2	
		4, 3, 2, 1	

Let's group the permutations from Table 6.1 according to the value of their anchor, see Table 6.2. Denote by x_i the number of permutations in the column $r = i$. We have $x_1 = 0$, $x_2 = 8$, $x_3 = 10$, $x_4 = 6$.

Definition 6.3 Let x_i be the total number of permutations such that their anchor equals i, $i = 1, 2, \ldots, n$. The set

$$D = \left\{ d_1 = \frac{x_1}{n!}, \ldots, d_n = \frac{x_n}{n!} \right\} \tag{6.2.1}$$

is called the *destruction spectrum* of the network. We will use the abbreviation *D-spectrum*.

The D-spectrum for our network is

$$D = \left\{ 0, \frac{1}{3}, \frac{5}{12}, \frac{1}{4} \right\}.$$

The values d_i give the discrete density of the network anchor value and therefore $d_1 + d_2 + \cdots + d_n = 1$.

The D-spectrum defined by (6.2.1) coincides with Internal Distribution [1] and signature [2].

It will be not superfluous to repeat that the D-spectrum *depends only on the network structure and its UP/DOWN criteria*, and is not related to the *up/down* probabilities of the network elements.

Definition 6.4 Let $y_b = \sum_{i=1}^{b} d_i$, $b = 1, \ldots, n$. Then the set $(y_1, \ldots, y_n)$ is called the *cumulative D-spectrum or CD-spectrum*.

As we noted in Definition 6.1, fundamental property of the CD-spectrum is the relationship between y_i and the number $C(i)$ of cut sets of the network of size i, given in the following theorem.

Theorem 6.1

$$C(i) = y_i \cdot \frac{n!}{i!(n-i)!},\qquad (6.2.2)$$

where $C(i)$ is the total number of cut sets of size i in the network.

We omit the proof of this statement. It can be found in [3, 4].

The following important theorem gives the network reliability formula using CD-spectrum.

Theorem 6.2 *If all $p_i = p$, then network reliability $R(N)$ can be expressed in the following form:*

$$R(N) = 1 - \sum_{i=1}^{n} y_i \cdot \frac{n!q^i p^{n-i}}{i!(n-i)!}\qquad (6.2.3)$$

Theorem 6.2 is easily obtained from Theorem 6.1.

Indeed, the probability of each cut set of size i is equal to $q^i p^{n-i}$. Multiplying (6.2.2) by this probability and then summing by i, we arrive at the network $DOWN$ probability, and from that we get (6.2.3).

Continuing our example, we get that the CD-spectrum for our network is: ($y_1 = 0$, $y_2 = \frac{1}{3}$, $y_3 = \frac{3}{4}$, $y_4 = 1$). By the formula 6.2.3 we get:

$$R = 1 - \left(\frac{1}{3} \cdot \frac{4!q^2 p^2}{2!2!} + \frac{3}{4} \cdot \frac{4!q^3 p}{3!1!} + 1 \cdot \frac{4!q^4}{4!} \right) = 1 - 2q^2 + q^3.$$

Remark 6.1 An alternative way of obtaining the important formula (6.2.3) is based on using order statistics. Let us return to Sect. 1.4 and consider again formula (1.4.2):

$$P(\tau \le t) = \sum_{k=1}^{k=n} \sum_{j=1}^{k} f_j \frac{n!}{k!(n-k)!}[G(t)]^k[1 - G(t)]^{n-k}.\qquad (6.2.4)$$

The f_j values, by their definition, coincide with D-spectrum (or signature). Therefore $f_1 + f_2 + \cdots + f_k = y_k, k = 1, 2, \ldots, n$ is the CD-spectrum. Now note that $P(\tau \le t) = P(DOWN), G(t) = q, 1 - G(t) = 1 - q = p$. Then (6.2.4) after simple algebra will coincide with (6.2.3).

Table 6.3 shows the reliability of the dodecahedron network from Fig. 5.2. For our calculations, we define the nodes as unreliable, and the edges as reliable. The calculations were performed by two methods: using the Destruction Monte Carlo (DMC) with $M = 10,000$ and using CMC with $M = 10,000$ and $M = 100,000$. The relative error for both algorithms was calculated in accordance with 4.3.4-4.3.5, with $K = 10$.

When $M = 10,000$ for both algorithms, the relative error of DMC is lower than CMC. Moreover, for node up probability 0.99, we see the presence of the rare event

Table 6.3 Dodecahedron network reliability by destruction Monte Carlo (DMC) and CMC

Algorithm/p	0.8	0.95	0.99
DMC, M = 10,000. $\hat{R}$	0.9533	0.99954	0.9999967
r.e.($\hat{Q}$)	1.2%	7.1%	12.4%
CMC, M = 10,000. $\hat{R}$	0.9539	0.99951	1
r.e.($\hat{Q}$)	5.1%	42.4%	Undefined
CMC, M = 100,000. $\hat{R}$	0.9531	0.99956	0.999998
r.e.($\hat{Q}$)	1.2%	13.8%	316%

phenomenon in CMC: in all 10,000 iterations the network state was UP. In the case of CMC with $M = 100,000$, the relative error of DMC with $M = 10,000$ remains lower.

6.3 BIM Spectrum

Theorem 6.2 from the previous section gives an expression of network reliability using the CD-spectrum. In this section, we define a new notion—the **BIM-spectrum**, and give an expression of BIM of an element using this spectrum.

The following explanation is useful for a better understanding of the structure of the spectrum. A reader interested in the practical side of the topic may refer directly to Definition 6.5.

Consider the network of Fig. 6.3. Let's ask the following question: what part in network destruction takes each network element. Consider all 8 permutations with the anchor equal 2 (Table 6.2):

$$(1, 4, 2, 3), (1, 4, 3, 2), (4, 1, 2, 3), (4, 1, 3, 2), (2, 4, 1, 3), (2, 4, 3, 1),$$
$$(4, 2, 1, 3), (4, 2, 3, 1)$$

Our network may collapse after two steps. The number of permutations in which node 1 is in the first or second place is equal to 4. By this number we will measure the participation of the node after two steps of destruction.

Consider now all permutations such that after **three** steps the network will be in $DOWN$. (Note that we do not say: for the first time $DOWN$.) It is clear that such permutations will be the permutations written above (with anchor 2), as well as all the following permutations with anchor 3.

$$(1, 2, 4, 3), (1, 3, 4, 2), (2, 1, 4, 3), (2, 3, 4, 1), (3, 1, 4, 2), (3, 2, 4, 1),$$
$$(3, 4, 1, 2), (3, 4, 2, 1), (4, 3, 1, 2), (4, 3, 2, 1)$$

Table 6.4 BIM's for network in Fig. 6.1

i	$z_{i,1}$	$z_{i,2}$	$z_{i,3}$	$z_{i,4}$
1	0	0	0	0
2	1/6	1/12	0	1/3
3	1/2	1/2	1/2	3/4
4	1	1	1	1

Participation of node 1 in the destruction of the network after 3 steps, we measure by the number of permutations of the above two groups, in which the node 1 is in the first, second or third position. We will obtain the number 12 (6 in the first group and also 6 in the second).

It is clear that after the fourth step, all permutations lead to the destruction of the network, i.e. the participation of the node 1 is 24.

The following is a formal definition of element participation in the destruction of the network in terms of the BIM-spectrum.

Definition 6.5 Denote by $Z_{i,j}$ the number of permutations satisfying the following two conditions:
(a) If the first i elements in the permutation are *down*, then the network is *DOWN*;
(b) Element j is among the first i elements of the permutation.
The collection of $z_{i,j} = Z_{i,j}/n!$ values, $i = 1, \ldots, n$; $j = 1, \ldots, n$ is called *BIM-spectrum* of the network. The set of $z_{i,j}$ values for fixed j and $i = 1, \ldots, n$ is called the BIM_j-spectrum, or the **importance spectrum** of element j.

Table 6.4 represents BIM-spectrum for all nodes of the network. The values $Z_{i,1}$ for node 1 were calculated above ($Z_{1,1} = 0$, $Z_{2,1} = 4$, $Z_{3,1} = 12$, $Z_{4,1} = 24$). So we get $z_{2,1} = 4/4! = 1/6$, $z_{3,1} = 12/4! = 1/2$. Similarly, we obtain values $Z_{i,j}$ for the remaining nodes $j = 2, 3, 4$ and calculate $z_{i,j}$.

Note that similar to the D-spectrum and CD-spectrum, BIM-spectrum is a **topological (structural) invariant**.

The following theorem allows us to calculate the Birnbaum importance defined in Chap. 5, using BIM-spectrum and CD-spectrum.

Theorem 6.3 BIM_j, $j = 1, \ldots, n$, equals

$$BIM_j = \sum_{i=1}^{n} \frac{n!(z_{i,j} \cdot q^{i-1} p^{n-i} - (y_i - z_{i,j}) q^i p^{n-i-1}}{i!(n-i)!} \tag{6.3.1}$$

Note that $y_n - z_{n,j} = 0$. This means that in the second term of the numerator, it is possible to assume that index i changes from 1 to $n - 1$.
The proof of this theorem the reader will find given in [3].

Continuing our example, we get using (6.3.1) for node 1: $BIM_1 = q - q^2$, for all q. It is easy to verify that a calculation using formula (5.1.4) gives exactly the same result.

In some cases, it is possible to order elements by their importance without calculating the corresponding $BIM's$. Such cases are described by the following theorem.

Theorem 6.4 *Suppose we are given the $BIM's$ for the network. Let us fix two indices α and β, $\alpha \neq \beta$, and the corresponding $Z_{i,\alpha}$ and $Z_{i,\beta}$ values. Then*
(1) If for all i, $i = 1, \ldots, k$, $Z_{i,\alpha} \geq Z_{i,\beta}$, then $BIM_\alpha \geq BIM_\beta$ for all p values.
(2) If at least one index i satisfies the exact inequality $Z_{i,\alpha} > Z_{i,\beta}$, then $BIM_\alpha > BIM_\beta$ for all p values.
(3) Suppose that (1) does not take place, and let m be the maximal index such that $Z_{m,\alpha} \neq Z_{m,\beta}$. Without loss of generality, assume that $Z_{m,\alpha} > Z_{m,\beta}$. Then there exists p_0 such that for $p \geq p_0$, $BIM_\alpha > BIM_\beta$.

Remark 6.2 It is clear that the Theorem 6.4 admits the use of $z_{i,\alpha}$ and $z_{i,\beta}$ instead of $Z_{i,\alpha}$ and $Z_{i,\beta}$.

Using this theorem, let us compare the spectra of the nodes in Table 6.4. We will obtain that

$$BIM_4 > BIM_1 > BIM_2 > BIM_3,$$

for all p values.

6.4 CD-Spectrum and BIM Spectra Monte Carlo

The exact calculation of network CD-spectrum and BIM-spectra is a formidable task. We suggest estimating the spectra using Monte Carlo approach. Below we present an algorithm which simultaneously estimates the CD-spectrum and the BIM-spectra for all network elements.

Algorithm 6.1: Evaluation of CD-spectrum and BIM-spectra.
1. Initialize all a_i and $b_{i,j}$ to be zero, $i = 1, \ldots, n$; $j = 1, \ldots, n$.
2. Simulate permutation π of all elements.
3. Find out the *anchor* $r(\pi)$. // See Remark 6.3 for an explanation of anchor search.
4. Put $a_r = a_r + 1$.
5. Find all j such that element j occupies one of the first r positions in π, and for each such j **Put** $b_{r,j} = b_{r,j} + 1$.
6. Put $r = r + 1$. **If** $r \leq n$ **GOTO 4**.
7. Repeat 2–6 M times.
8. Estimate $y_i, z_{i,j}$ via $\hat{y}_i = \frac{a_i}{M}$, $\hat{z}_{i,j} = \frac{b_{i,j}}{M}$.

Remark 6.3 The position of the anchor can be *efficiently* found by applying **bisectional** algorithm which works as follows.

Erase the $\lfloor n/2 \rfloor$ elements of the permutation. **Check** the network state. **If** it is already $DOWN$, the anchor must be in the first $\lfloor n/2 \rfloor$ positions. **If** not, the anchor is within remaining positions. **Proceed** by bisecting the part of the permutation until you locate the anchor.

Table 6.5 Dodecahedron BIM-spectrum. Nodes unreliable. Terminals $T = (1, 14, 19)$

i	$z_{i,2}$	$z_{i,3}$	$z_{i,4}$	$z_{i,5}$	$z_{i,6}$	$z_{i,7}$	$z_{i,8}$	$z_{i,9}$
3	0.0016	0	0.0013	0.0016	0	0.0	0.0016	0.0
4	0.0082	0.0021	0.0082	0.0086	0.0022	0.0018	0.0081	0.0021
5	0.027	0.013	0.029	0.029	0.012	0.012	0.026	0.015
6	0.080	0.060	0.083	0.085	0.052	0.050	0.073	0.057
7	0.18	0.15	0.19	0.19	0.14	0.14	0.18	0.15
8	0.31	0.28	0.32	0.33	0.28	0.28	0.30	0.28
9	0.44	0.42	0.46	0.45	0.42	0.41	0.43	0.43
10	0.55	0.54	0.56	0.56	0.54	0.54	0.54	0.55
11	0.63	0.63	0.64	0.64	0.63	0.63	0.62	0.63

On the average, the location of the anchor will be found by $O(log_2 n)$ network state checks.

Below is shown the CD-spectrum of the dodecahedron (Fig. 5.2) obtained by the above Monte Carlo algorithm.

$$y_1 = y_2 = 0, \ y_3 = 0.0042, \ y_4 = 0.022, \ y_5 = 0.07,$$
$$y_6 = 0.19, \ y_7 = 0.4, \ y_8 = 0.64, \ y_9 = 0.82,$$
$$y_{10} = 0.93, \ y_{11} = 0.98, \ y_{12} = 0.996, \ y_{13} = 0.999, \ y_{14} = y_{15} = y_{16} = y_{17} = 1.$$

BIM-spectra for dodecahedron nodes 2–9 are presented in Table 6.5.

The first two rows are zeros, since the minimum cut of the network is 3. Spectrum values in the range of 12–17 are not shown. These values are almost the same, since the probability of network failure starting from step 12 is very close to 1, and at step 15 is already equal to 1.

By comparing the spectra, one can rank the vertices by their importance as follows:

$$group1(node4 = node5 = node13 = node20) >$$
$$group2(node2 = node8 = node11 = node15 = node18) >$$
$$group3(node3 = node6 = node7 = node9 = node10 = node12 = node16 = node17)$$

Note that the group of best nodes coincides with the group obtained using CMC in 5.3.

Spectra for nodes from the first and third groups are presented in Fig. 6.4.

CD-Spectra and Resilience of the Ring and Grid

Let us now return to the networks in Figs. 6.1 and 6.2, introduced in Sect. 6.1. Our intention was to compare the behaviour of these networks in the "shock" process of

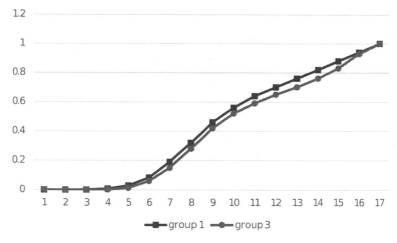

Fig. 6.4 Dodecahedron BIM-spectrum

Table 6.6 Comparing CD-spectra of the ring and grid for terminal connectivity

i	1	2	3	4	5	6	7	8
Ring	0	0.005	0.026	0.077	0.158	0.282	0.421	0.585
Grid	0	0.014	0.047	0.104	0.198	0.328	0.494	0.674
i	9	10	11	12	13	14	15	16
Ring	0.746	0.862	0.940	0.979	0.994	0.998	0.9998	1
Grid	0.831	0.926	0.974	0.992	0.998	0.9999	1	

Table 6.7 Comparing CD-spectra of the ring and grid for all nodes connectivity

i	1	2	3	4	5	6	7	8	9
Ring	0	0.034	0.117	0.244	0.422	0.617	0.793	0.933	1
Grid	0	0.015	0.060	0.148	0.302	0.499	0.717	0.904	1

successive random removal of edges. We will use for this purpose Tables 6.6, 6.7, 6.8 which give the spectra for various criteria.

Consider Table 6.6 and the corresponding graph in Fig. 6.5. We see that for each i the value of the CD-spectrum of the ring is **less** than the corresponding value for the grid. This means that the ring disintegrates with smaller probability at each step of the destruction process. We emphasize once again that this conclusion does not depend on the *up/down* edge probabilities.

We see from Table 6.7 and the corresponding graph in Fig. 6.6 that in the case of all nodes connectivity, the ring decays at each step with **greater** probability than the grid.

Table 6.8 Comparing CD-spectra of the ring and grid for Maximal cluster criterion ($maxsize \geq 10$)

i	1	2	3	4	5	6	7	8
Ring	0	0	0	0.0012	0.0075	0.023	0.060	0.137
Grid	0	0	0.0018	0.0056	0.015	0.035	0.076	0.142
i	9	10	11	12	13	14		
Ring	0.269	0.453	0.664	0.855	0.966	1		
Grid	0.253	0.417	0.621	0.829	0.956	1		

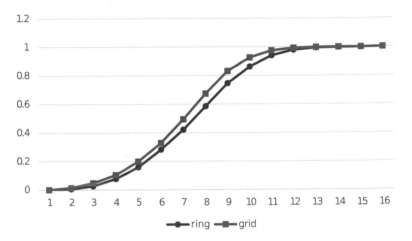

Fig. 6.5 Comparing CD-spectra for terminal connectivity

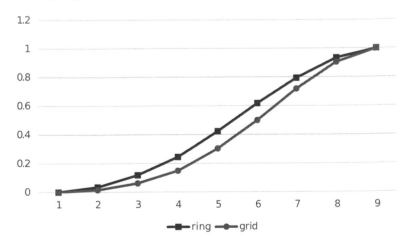

Fig. 6.6 Comparing CD-spectra for all nodes connectivity

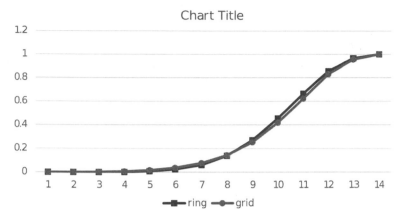

Fig. 6.7   Comparing CD-spectra for maximal cluster criterion

In the case of the maximum cluster criterion (Table 6.8 and Fig. 6.7), up to step 8, the grid disintegrates with greater probability, and starting from step 9—the ring becomes less reliable.

Consider now a comparison of networks in terms of resilience. In fact, the concepts of resilience and CD-spectrum are closely related. Indeed, recall the definition of resilience (Definition 3.3):

The probabilistic resilience $res_{pr}(N; \beta)$ is the largest number of element failures such that N is *DOWN* with probability less then β.

Thus, from the CD-spectrum, we can get the network resilience for any β. Let, for example, $\beta = 0.03$. Then we get:

For terminal connectivity: $res_{pr}(\textbf{ring}; 0.03) = 3 \; res_{pr}(\textbf{grid}; 0.03) = 2$
For all terminal connectivity: $res_{pr}(\textbf{ring}; 0.03) = 1 \; res_{pr}(\textbf{grid}; 0.03) = 2$
For maximal cluster: $res_{pr}(\textbf{ring}; 0.03) = 6 \; res_{pr}(\textbf{grid}; 0.03) = 5.$

References

1. Elperin, T., Gertsbakh, I. B., & Lomonosov, M. (1991). Estimation of network reliability using graph evolution models. *IEEE Transactions on Reliability, 40*(5), 572–581.
2. Samaniego, F. G. (2007). *System signatures and their applications in engineering reliability.* New York: Springer.
3. Gertsbakh, I., & Shpungin, Y. (2009). *Models of network reliability: Analysis, combinatorics and Monte Carlo.* Boca Raton: CRC Press.
4. Gertsbakh, I., & Shpungin, Y. (2011). *Network reliability and resilience.* New York: Springer.

Chapter 7
Lomonosov's Turnip

Abstract Lomonosov's turnip is a very powerful Monte Carlo algorithm for estimating Network reliability. It was invented by M.V. Lomonosov in 1986 and first
described in [1]. This algorithm is applicable to finding the probability of network
terminal (or all-node) connectivity probability for the case of arbitrary values of
edge (or node) failure probabilities. An outstanding feature of this algorithm is that it
enables obtaining estimates with high accuracy for very small network failure probabilities known as "**rare event** situation". The name "turnip" comes from a diagram
reminding a turnip which describes so-called *evolution process* which is central to
the main idea of the algorithm.

Keywords Evolution process · Turnip diagram · Closure · Edge and node failure

7.1 Evolution Process

Let us introduce a time axis and consider a "birth" process for the family of all edges
of the network. This process starts at time $t = 0$. Edge e is born at random moment
$\xi(e)$ which is exponentially distributed with parameter $\lambda(e)$. Before this moment the
edge does not exist and after its birth, remains *up* forever. So,

$$P(\xi(e) \leq t) = 1 - exp[-\lambda(e) \cdot t)], e \in E$$

Let us choose an arbitrary moment of time t, say $t = t_0 = 1$ and for each edge e take
$\lambda(e)$ such that,

$$e^{-\lambda(e)} = q(e) = P(\xi(e) > 1).$$

Now imagine that you observe the above evolution process at the instant $t_0 = 1$.
You will see the edge e **born** if it is born before $t_0 = 1$, that is with probability
$p(e) = 1 - q(e)$. If the edge e is born after $t_0 = 1$, it will **not exist** at this instant
and this will happen with probability $q(e)$.

© The Author(s), under exclusive license to Springer Nature Singapore Pte Ltd. 2020   71
I. Gertsbakh and Y. Shpungin, *Network Reliability*,
SpringerBriefs in Electrical and Computer Engineering,
https://doi.org/10.1007/978-981-15-1458-6_7

Fig. 7.1 Network with 5
nodes and 6 edges

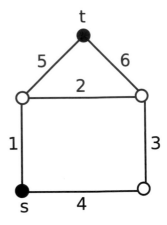

Summing up, the "instant photography" of the evolution process taken at $t_0 = 1$ is exactly the same as the result of a random lottery in which each edge e is *up* with probability $p(e)$ and *down* with probability $q(e) = 1 - p(e)$.

Now let us take a particular network and investigate what will happen if all its edges will be born according with the above evolution/birth process.

Our example will be a small network with five nodes and six edges, see Fig 7.1 below. The elements subject to failure are the edges. Edge i has birth rate λ_i, $i = 1, 2, 3, 4, 5, 6$. Two nodes, s and t are **terminals** and network failure is defined as the loss of connection between these terminals.

We will need a small theorem which we present without proof:

Theorem 7.1 *Suppose we have a family of k "objects" which have independent exponentially distributed birth times. In a birth process starting at $t = 0$, object j has birth time $\tau_j \sim Exp(\lambda_j)$, $j = 1, 2, \ldots, k$.*

Then the first (the earliest) birth will be at the instant τ which has exponential distribution with parameter $\Lambda = \lambda_1 + \cdots + \lambda_k$, and with probability

$$\alpha_j = \frac{\lambda_j}{\Lambda}$$

the first (the earliest) birth will belong to the jth object.#

Let us proceed now to investigate the network behaviour in its evolution process.

The diagram on Fig. 7.2 shows the evolution process going on our network edges. The lowest circle on Fig. 7.2 (the "root") presents the situation when no edge is born, say at $t = 0$. We call it superstate σ_0. Second "floor" of the turnip represents six different possibilities of the first (the earliest) birth, and we show only four of them. For example, the first circle from the right shows the "superstate" $\sigma_{1,4}$ when the first birth was of edge 5. According to the above theorem, the transition $\sigma_0 \to \sigma_{1,4}$ happens with probability $\lambda_5/(\lambda_1 + \lambda_2 + \cdots + \lambda_6)$. The duration of this transition is $\tau_0 \sim Exp(\Lambda)$, where $\Lambda = \lambda_1 + \cdots + \lambda_6$.

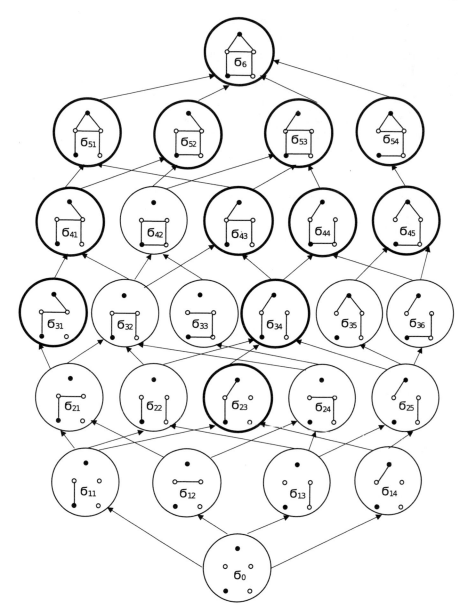

Fig. 7.2 The turnip diagram for network on Fig. 7.1

Third "floor" shows all superstates with 2 edges born. There are $6!/(2!4!) = 15$ superstates on third floor and we show only 5 of them. Only one of the superstates represents network *UP* state (see double circled superstate $\sigma_{2,3}$). Here edges 1 and 5 create a path from s to t. The transition from $\sigma_{1,1}$ to $\sigma_{2,3}$ takes place with probability $\lambda_5/(\lambda_2 + \lambda_3 + \lambda_4 + \lambda_5 + \lambda_6)$.

Fourth level of the turnip has $6!/(3!3!) = 20$ superstates. We show only 6 of them. Several superstates (double circled) represent UP states.

The fifth level of the turnip shows superstates with four up edges. In total, there are $6!/(4!2!) = 15$ such superstates and only 5 of them are shown. Most of the superstates of this level correspond to network UP state.

The sixth level represent superstates with 5 up edges. All six such superstates represent the network in state UP, and we show four of them. The last seventh level represents superstate with all 6 edges in up.

Let us now give an important definition of a *trajectory*

Definition 7.1 A trajectory is a sequence $u = (\sigma_0, \sigma_{1,i_1}, \sigma_{2,i_2}, \ldots, \sigma_{r,i_r})$ of superstates such that the initial superstate is the root σ_0, each σ_{j,i_j} is a direct successor of $\sigma_{j-1,i_{j-1}}$, and the last superstate is network UP state σ_{r,i_r}. #

Remark 7.1 Each direct successor is obtained by the earliest birth of one edge of the previous super-state. For example, $\sigma_0, \sigma_{1,1}, \sigma_{2,2}, \sigma_{3,4}$ is a trajectory, and $\sigma_{3,4}$ is a direct successor of $\sigma_{2,2}$.

Denote by $p(u)$ the probability that the evolution process goes along trajectory u. It is easy to calculate this probability:

$$p(u) = \Pi_{j=0}^{r-1} P(\sigma_{j,i_j} \to \sigma_{j+1,i_{j+1}}), \tag{7.1.1}$$

where $\sigma_{0,i_0} = \sigma_0$.

For example, let us compute the probability of the trajectory $u = \sigma_0, \sigma_{1,1}, \sigma_{2,2}, \sigma_{3,4}$:

$$p(u) = \frac{\lambda_1}{\sum_{i=1}^{6} \lambda_i} \cdot \frac{\lambda_3}{\sum_{i=2}^{6} \lambda_i} \cdot \frac{\lambda_5}{\lambda_2 + \lambda_4 + \lambda_5 + \lambda_6}.$$

7.2 Lomonosov's Algorithm

Simulating the trajectory leading from the root of the turnip to the UP state is the first step in applying Lomonosov's algorithm [2, 3].

The second step is **determining the time** of the transition from the root to the UP state. We know that the evolution process stays in a superstate $\sigma_{i,j}$ random time $\tau(\sigma_{i,j})$ which is exponentially distributed with parameter $\Lambda(\sigma_{i,j})$.

The total time along the evolution trajectory is a sum of independent exponentially distributed random variables. More formally, if the trajectory is $u = \sigma_0 \to \sigma_{1,i_1} \to \cdots \to \sigma_{r,i_r}$, then

$$P(t|u) = P[\tau(\sigma_0) + \tau(\sigma_{1,i_1}) + \cdots + \tau(\sigma_{r,i_r}) \leq t|u] \tag{7.2.1}$$

Let us note that $P(t|u)$ is a convolution of several exponentials and can be easily computed by a formula given in Sect. 7.4.

Now we are ready to write the principal formula for the Lomonosov's algorithm. At $t = 0$ the network is $DOWN$ (there are no edges). Denote by $\xi(\mathbf{N})$ network's "birth time", i.e. the time when the network becomes UP. $P(\xi(\mathbf{N}) \leq t_0 = 1)$ is the probability that at $t_0 = 1$ the network is UP, i.e "was born" in the evolution process.

Theorem 7.2 *By the Law of the total probability,*

$$R(N) = P(\xi(\mathbf{N}) \leq t_0 = 1)) = \Sigma_{u \in U} P(u) P(t|u) \qquad (7.2.2)$$

where U is the set of all trajectories leading from σ_0 to the UP state.#

The expression (7.2.2) has a form of a mean value since $\Sigma_{u \in U} P(u) = 1$, and is the key for estimating $R(N)$ by means of the following Monte Carlo algorithm.

Algorithm 7.1
1. Put $\hat{R} = 0$
2. Generate trajectory u leading from σ_0 to the super-state in UP. **Use** for its generation the above described transition probabilities.
3. Calculate $\hat{R} = \hat{R} + P(t|u)$.
4. Repeat 2 and 3 M times.
5. Put $\hat{R}(N) = \frac{\hat{R}}{M}$.

Remark 7.2 The trajectory u is has probability (7.1.1) which *does not depend* on parameter t. This explains why in the turnip scheme the "rare event" phenomenon does not exist [3, 4].

Remark 7.3 As in Algorithm 6.1, the efficiency of this algorithm can be improved using the bisectional method. Here we are dealing with a birth (construction) process, and therefore we use a slightly modified definition of the anchor.

The ordinal number of the first element in the birth process, when the network becomes UP, is called the *anchor* of the trajectory (permutation).
So the procedure of bisection will be as follows.
Insert $\lfloor n/2 \rfloor$ elements of the permutation. **Check** the network state. **If** it is already UP, the anchor must be in the first $\lfloor n/2 \rfloor$ positions. **If** not, the anchor is within remaining positions. **Proceed** by bisecting the part of the permutation until you locate the anchor.

Remark 7.4 The evolutionary process outlined in the previous section and Lomonosov's algorithm in this section are simplified. In the original version one of the significant components of this process is the operation of **closure** [2, 3], which allows to increase the efficiency of the turnip algorithm.

Various aspects of the efficiency of the Lomonosov's algorithm were considered in [3, 4]. In simple terms, this algorithm is highly efficient, it avoids the so-called rare event phenomenon, and has several advantages over the Crude Monte Carlo algorithm. One

Table 7.1 Dodecahedron network reliability by turnip and CMC algorithms

Algorithm/p	0.8 for e_1-e_{15} 0.9 for $e_{16}-e_{30}$	0.9 for e_1-e_{15} 0.99 for $e_{16}-e_{30}$	0.99 for all edges
Turnip, M = 10,000. $\hat{R}$	0.9796	0.99865	0.9999978
r.e.($\hat{Q} = 1 - \hat{R}$)	4.1%	8.4%	43.7%
CMC, M = 10,000. $\hat{R}$	0.9784	0.9979	1
r.e.($\hat{Q} = 1 - \hat{R}$)	8.1%	22.7%	Undefined
CMC, M = 100,000. $\hat{R}$	0.9798	0.99859	0.99999
r.e.($\hat{Q} = 1 - \hat{R}$)	1.9%	11.4%	316%

of these advantages is the possibility, using the algorithm for computing the vector of BIM's for all elements, the so-called **reliability gradient** [3, 5].

We will limit ourselves to an example that compares the behaviour of the relative error of the turnip algorithm and the CMC.

Table 7.1 shows the reliability of the dodecahedron network from Fig. 5.2. For our calculations, we define *edges* as unreliable, and nodes as reliable. The UP criterion is three terminals connectivity. The calculations were performed by two methods: using the turnip algorithm with $M = 10,000$ and using CMC with $M = 10,000$ and $M = 100,000$. The relative error for both algorithms was calculated in accordance with 4.3.4-4.3.5, with $K = 10$.

We see from this table that when $M = 10,000$ for both algorithms, the relative error of turnip is lower than of CMC. Moreover, for edge *up* probability 0.99, we see the presence of the rare event phenomenon in CMC: in all 10,000 iterations the network state was UP. When $M = 100,000$ for CMC, only in the second column we see that r.e. is lower for CMC, whereas for higher probabilities, the relative error of Lomonosov's algorithm with $M = 10,000$ remains lower than that of CMC with $M = 100,000$.

7.3 Turnip for Networks with Unreliable Nodes

In this section we assume that the edges are always up, and the nodes, except the terminal nodes, are subject to failures and fail independently [6]. Remind that node failure means that all edges incident to this node are erased and the failed node becomes isolated. If nodes i and j both are up and there exists an edge $e(i, j)$ connecting these nodes, then edge $e(i, j)$ becomes operational (up). If one of two nodes that are connected by some edge are down, then the corresponding edge is assumed to be non existing (erased).

The *birth* process on the nodes develops in the following way. First, we have an initial super-state σ_0 which contains only terminal nodes and no nodes which are subject to failure, see Fig. 7.4. We assume that there are no direct edges between

Fig. 7.3 Network with 6 nodes and 8 edges. Nodes are unreliable

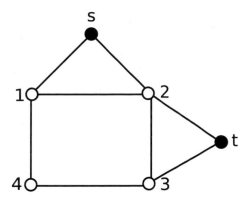

terminal nodes. Then nodes are born at random instants. As soon as two nodes a and b are born, and there is an edge $e = (a, b)$ in our network, these nodes get immediately connected by this edge. Similarly, if there is an edge connecting a terminal node to a newborn node, say a, this edge immediately becomes "alive". The node birth and evolution process goes exactly as the above described process on the edges. As to the edges, they appear automatically as soon as their end-point nodes are born.

All the above introduced notions—birth rates, trajectories, super-states, transition rates, sitting times—remain with no changes. All the theory and the simulation algorithm which were described for edge evolution process remain valid for node evolution process.

Consequently, all theoretical results and algorithms can be translated in equivalent form from the "edge language" into the "node language" [6].

Consider a network with six nodes and eight edges shown on Fig. 7.3. The elements subject to failure are the nodes. Two nodes, s and t are terminals. The UP criterion is terminal connectivity. Assume that node i has birth rate λ_i, $i = 1, 2, 3, 4$.

The diagram on Fig. 7.4 shows the evolution process going on our network. As noted above, the evolution process on the nodes goes exactly as on the edges. Consider, for example, the trajectory $\sigma_0 \to \sigma_{1,4} \to \sigma_{2,3} \to \sigma_{3,2}$. The first transition from σ_0 to $\sigma_{1,4}$ takes place with probability $\lambda_4/(\lambda_1 + \lambda_2 + \lambda_3 + \lambda_4)$. For the second transition $\sigma_{1,4} \to \sigma_{2,3}$, the probability is $\lambda_1/(\lambda_1 + \lambda_2 + \lambda_3)$, the last transition takes place with probability $\lambda_2/(\lambda_2 + \lambda_3)$. Finally, the probability of the entire trajectory is calculated as the product of the probabilities of all transitions:

$$p(u) = \frac{\lambda_4}{\sum_{i=1}^{4} \lambda_i} \cdot \frac{\lambda_1}{\sum_{i=1}^{3} \lambda_i} \cdot \frac{\lambda_2}{\lambda_2 + \lambda_3}$$

Table 7.2 shows the reliability of the dodecahedron network from Fig. 5.2. For our calculations, we define the nodes as unreliable, and the edges as reliable. The calculations were performed by two methods: using the turnip algorithm with $M = 10,000$ and using CMC with $M = 10,000$ and $M = 100,000$. The relative error for both algorithms was calculated in accordance with 4.3.4-4.3.5, with $K = 10$.

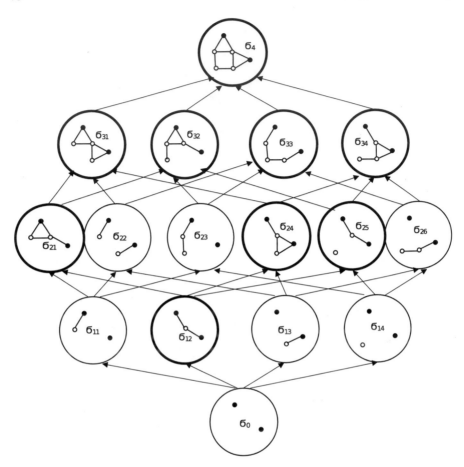

Fig. 7.4 The turnip diagram for network on Fig. 7.3

Table 7.2 Dodecahedron network reliability by CMC and turnip algorithm

Algorithm/p	0.8 for nodes 2–9 0.9 for other	0.9 for nodes 2–9 0.99 for other	0.99 for all nodes
Turnip, M = 10,000. $\hat{R}$	0.9819	0.99878	0.999998
r.e.($\hat{Q}$)	4.1%	4.4%	15.7%
CMC, M = 10,000. $\hat{R}$	0.9793	0.9986	1
r.e.($\hat{Q}$)	8.3%	34.3%	Undefined
CMC, M = 100,000. $\hat{R}$	0.9821	0.99863	0.99999
r.e.($\hat{Q}$)	1.5%	7.6%	105.4%

Here we see the same situation as in Table 7.1 for the case of unreliable edges. Let us repeat this conclusion also for the case of unreliable nodes.

When $M = 10,000$ for both algorithms, the relative error of turnip is lower than CMC. Moreover, for node up probability 0.99, we see the presence of the rare event phenomenon in CMC: in all 10,000 iterations the network state was UP. When $M = 100,000$ for CMC, only in the first column we see that r.e. is lower for CMC, whereas for higher probabilities, the relative error of Lomonosov's algorithm with $M = 10,000$ remains lower than that of CMC with $M = 100,000$.

7.4 Convolution of Exponents

References [3, 7].

In this section we will consider the exact analytic expression for the convolution of r exponential distributions with parameters Λ_i, $i = 0, \ldots, r - 1$. It will be assumed that

$$\Lambda_0 > \Lambda_1 > \Lambda_2 > \ldots > \Lambda_{r-1}. \tag{7.4.1}$$

The fact that we take without proof is the following: If $\tau_i \sim Exp(\Lambda_i)$, $i = 0, 1, \ldots, r - 1$, then

$$P(\tau_0 + \tau_1 + \cdots + \tau_{r-1} \leq t) = 1 - \sum_{k=1}^{r} A_{r,k} e^{-\Lambda_{k-1} t}, \tag{7.4.2}$$

and

$$\sum_{k=1}^{r} A_{r,k} = 1. \tag{7.4.3}$$

The last formula follows from the fact the the convolution equals zero when we substitute $t = 0$ into (7.4.2).

The coefficients $A_{r,k}$ are found by the following recursive procedure.

$$A_{1,1} = 1; \tag{7.4.4}$$

$$A_{r+1,k} = A_{r,k} \cdot \frac{\Lambda_r}{\Lambda_r - \Lambda_{k-1}}, \quad k = 1, 2, \ldots, r;$$

$$A_{r+1,r+1} = 1 - \sum_{k=1}^{r} A_{r+1,k}.$$

Example Let us consider a convolution of $r = 4$ exponents with parameters $\Lambda_0 = 4$, $\Lambda_1 = 3$, $\Lambda_2 = 2$, $\Lambda_3 = 1$.
First,
$$P(\tau_0 \leq t) = 1 - e^{\Lambda_0 t}.$$

Next

$$P(\tau_0 + \tau_1 \leq t) = 1 - A_{2,1}e^{-4t} - A_{2,2}e^{-3t} = 1 - A_{2,1}e^{-\Lambda_0 t} - A_{2,2}e^{-\Lambda_1 t}.$$

By (7.4.4), $A_{2,1} = A_{1,1}\Lambda_1/(\Lambda_1 - \Lambda_0) = -3$, $A_{2,2} = 1 - A_{1,1} = 4$.

So, the convolution of the first two exponents has the expression:

$$P(\tau_0 + \tau_1 \leq t) = 1 + 3e^{-\Lambda_0 t} - 4e^{-\Lambda_1 t}. \tag{7.4.5}$$

Now $P(\tau_0 + \tau_1 + \tau_2 \leq t) = 1 - A_{3,1}e^{-4t} - A_{3,2}e^{-3t} - A_{3,1}e^{-2t}$ and
$A_{3,1} = A_{2,1}\Lambda_2/(\Lambda_2 - \Lambda_0) = 3$ and
$A_{3,2} = A_{2,2}\Lambda_2/(\Lambda_2 - \Lambda_1) = -8$,
$A_{3,3} = 1 - A_{3,1} - A_{3,2} = 6$.

So, the convolution of the first three exponents has the expression

$$P(\tau_0 + \tau_1 + \tau_2 \leq t) = 1 + -e^{-\Lambda_0 t} - 8e^{-\Lambda_1 t} - 6e^{-\Lambda_2 t}. \tag{7.4.6}$$

Finally

$$P(\tau_0 + \tau_1 + \tau_2 + \tau_3 \leq t) = 1 - A_{4,1}e^{-4t} - A_{4,2}e^{-3t} - A_{4,3}e^{-2t} - A_{4,4}e^{-t}.$$

$A_{4,1} = A_{3,1}\Lambda_3/(\Lambda_3 - \Lambda_0) = -1$,
$A_{4,2} = A_{3,2}\Lambda_3/(\Lambda_3 - \Lambda_1) = 4$,
$A_{4,3} = A_{3,3}\Lambda_3/(\Lambda_3 - \Lambda_2) = -6$, and
$A_{4,4} = 1 - A_{4,1} - A_{4,2} - A_{4,3} = 4$.

Finally, the convolution of four exponents has the expression

$$P(\tau_0 + \cdots + \tau_3 \leq t) = 1 + e^{-\Lambda_0 t} - 4e^{-\Lambda_1 t} + 6e^{-\Lambda_2 t} - 4e^{-\Lambda_3 t}.\# \tag{7.4.7}$$

References

1. Lomonosov, M. (1994). On Monte Carlo estimates in network reliability. *Probability in the Engineering and Informational Sciences*, 8, 245–264.
2. Elperin, T., Gertsbakh, I. B., & Lomonosov, M. (1991). Estimation of network reliability using graph evolution models. *IEEE Transactions on Reliability*, 40(5), 572–581.
3. Gertsbakh, I., & Shpungin, Y. (2009). *Models of network reliability: Analysis, combinatorics and Monte Carlo*. Boca Raton: CRC Press.
4. Lomonosov, M., & Shpungin, Y. (1999). Combinatorics and reliability Monte Carlo. *Random Structures and Algorithms*, 14, 329–343.
5. Gertsbakh, I., & Shpungin, Y. (2011). *Network reliability and resilience*. New York: Springer.
6. Gertsbakh, I., Shpungin, Y., & Vaisman, R. (2014). Network reliability Monte Carlo with nodes subject to failure. *International Journal of Performability Engineering*, 10(2), 161–170.
7. Kroese, D., Taimre, T., & Botev, Z. I. (2011). *Handbook of Monte Carlo methods*. New York: Wiley.

Chapter 8
Examples of Network Analysis

Abstract This chapter presents three applications of the theory developed in previous chapters to practical analysis of network-type systems. The first section is a comparison of networks resilience under random attack on their nodes. It is shown that regular networks are considerably more resilient than randomly created networks having the same number of nodes and edges. The second section presents an example of predisaster reinforcement of highway/railway system based on locating and using the most influential (important) edges of the system. The third section presents reliability analysis of flow networks with randomly failing edges.

Keywords Regnet, ternet, prefnet · Predisaster reinforcement · Flow in random networks

8.1 Network Structure and Resilience Against Node Attack

Suppose we have a network which is subject to an attack on its nodes. The attacker chooses randomly a node in the network and destroys it. The attacked node becomes isolated and all edges incident to it are erased. This situation may reflect action of secret services aimed at the discovery and destruction of a terrorist network. Our first example is exactly this case and was borrowed from the report of Valdis Krebs published on his website www.orgnet.com. It describes the terrorist network in USA preparing their attack on 9/11/2001. We decided to compare the resilience of the terrorist network (we call it "ternet") to networks of approximately the same size but having different structure.

The ternet has been "created" in special circumstances, the nodes of it are the people, terrorists, and the edges—connections between them. Secret character of the whole situation made the connection quite random. We will compare the resilience of the ternet to the resilience of a more "organized", i.e. less randomly created network. We took two networks which we call regnet—regular network and prefnet—a network obtained by preferential assignment approach described in [1]. As a regular network we took a five dimensional cube H-5 with 32 nodes and 80 edges. Each node is incident to 5 edges and has degree 5. The nodes of regnet are 5-digit binary

© The Author(s), under exclusive license to Springer Nature Singapore Pte Ltd. 2020
I. Gertsbakh and Y. Shpungin, *Network Reliability*,
SpringerBriefs in Electrical and Computer Engineering,
https://doi.org/10.1007/978-981-15-1458-6_8

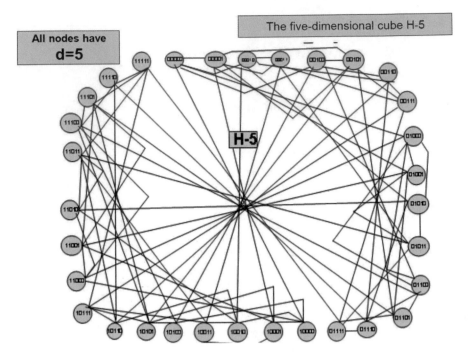

Fig. 8.1 Regnet

numbers from 00000 to 11111. If two nodes differ by only one binary digit, then they are connected by an edge. For example, nodes 00000 and 01000 are connected by an edge.

Prefnet is a network with the same number of 32 nodes and 80 edges. This network is made by having a small initial kernel of nodes and edges and adding new nodes and new edges following the principle: any new edge from a new node goes with higher probability to an existing node having more edges incident to it. This principle is called "preferential assignment" [1].

We designed our prefnet with 80 edges and 32 nodes. It has one node with 15 edges and one—with ten, and many nodes with degree 4–5. The average node degree is 5. The "real" ternet had 34 nodes and 91 edges. It has been modified randomly to a network with 32 nodes and 80 edges. The original ternet and the modified ternet have one hub with degree 16. Probably, it might correspond to the connections of the leader of the terrorist group (Muhammed Atta). Regnet, Prefnet and Ternet are shown on Figs. 8.1, 8.2 and 8.3 respectively.

So, we have a "most organized" Regnet, a "less organized" Prefnet and completely random "Ternet", all three having the same size and the same average degree.

Of crucial importance is the choice of the network *UP/DOWN* criterion. We assumed that the networks become *DOWN* if their *largest connected component becomes of size L ≤ 10*. We assume that any system designed to carry out a special

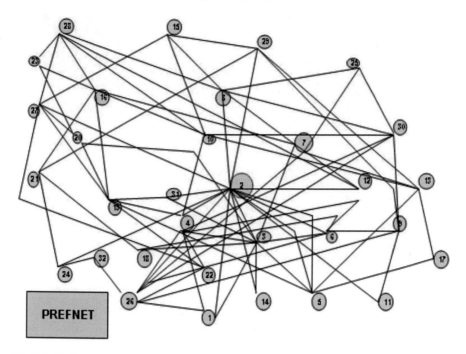

Fig. 8.2 Prefnet

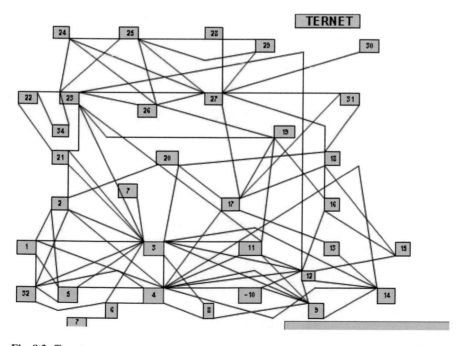

Fig. 8.3 Ternet

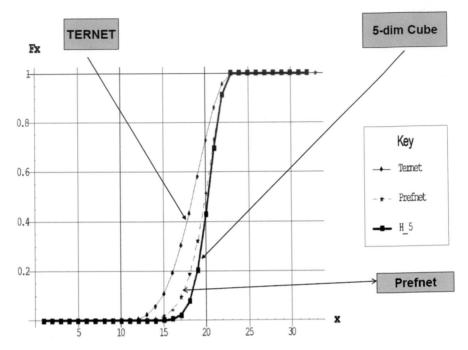

Fig. 8.4 CD-spectra of regnet, prefnet and ternet

mission is not more capable of doing so if it greatest **connected component** of the whole system is less than one third of the original number of nodes in the system.

The results of our analysis are presented on Fig. 8.4. It shows the CD-spectra for all three networks.

From Fig. 8.4 it becomes evident that the CD-spectrum $F(x)$ of Ternet dominates two other spectra and that the CD-spectrum of Prefnet dominates the CD-spectrum of Ternet. In other words, most resilient is Regnet and less resilient is Ternet. Compare, for example, the *DOWN* probabilities after failure of 17 nodes. For Ternet it is about 0.32, for Prefnet—0.1 and only 0.03 for Regnet.

8.2   Road/Highway Reinforcement

In this section we consider a transportation network reinforcement problem. By reinforcement we mean network reliability improvement achieved by replacing a certain number of its "weak" edges, by more reliable ones. If there are several ways to make the network "stronger", we will prefer the less costly way.

We show how this works taking as an example the ring way network with 15 nodes and 22 edges shown on Fig. 8.5.

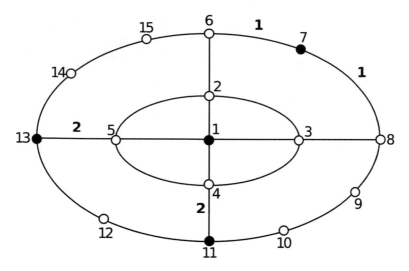

Fig. 8.5 Ring network. The network is *UP* if all four terminals (shown bold) are connected. Copy of Fig. 6.2

All edges have the same reliability $p(e) = 0.7$. Edge reinforcement means its replacement by more reliable one, say by an edge having $p = 0.9$. Suppose that edge replacement has the same cost for all edges.

The initial network reliability calculated using CMC is $R = 0.505$. Our goal is to raise this probability to a level $R^* = 0.85$, for minimal cost. The solution of our problem is equivalent to choosing the *minimal* number of edges to reinforce.

The analysis of BIM-spectra allows to divide the edges into the following six groups. (The edges belonging to the first and the second groups are marked in Fig. 8.5, by 1 and 2, respectively.)

$$(6, 7) = (7, 8) >$$
$$(4, 11) = (5, 13) >$$
$$(11, 12) = (12, 13) >$$
$$(2, 6) = (3, 8) >$$
$$(8, 9) = (9, 10) = (10, 11) = (13, 14) = (14, 15) = (6, 15) >$$
$$other$$

BIM-spectra of the edges, each of which represents a separate group, are shown in Table 8.1. The first row is zero, since the minimum cut of the network is 2. Spectrum values after $i = 12$ are not shown. These values are almost the same, since the probability of network failure starting from step 13 is very close to 1. Choosing for replacement one by one the edges with highest BIMs, we arrive at the reliability $R = 0.851$, using the following 6 edges:

Table 8.1 Ring BIM-spectrum. Edges unreliable. Terminals are $T = (1, 7, 11, 13)$

i	$z_{i,(1,2)}$ group 6	$z_{i,(2,6)}$ group 4	$z_{i,(4,11)}$ group 2	$z_{i,(6,7)}$ group 1	$z_{i,(8,9)}$ group 5	$z_{i,(11,12)}$ group 3
2	0	0	0	0.0042	0	0
3	0.0005	0.0026	0.0033	0.0145	0.0022	0.0035
4	0.0042	0.0141	0.0216	0.0362	0.0129	0.0193
5	0.0163	0.0442	0.0562	0.0732	0.0347	0.0501
6	0.0436	0.0898	0.1070	0.1200	0.0719	0.0950
7	0.0967	0.1532	0.1771	0.1824	0.1254	0.1579
8	0.1788	0.2257	0.2562	0.2574	0.1960	0.2368
9	0.2843	0.3138	0.3415	0.3436	0.2845	0.3210
10	0.3813	0.3939	0.4153	0.4100	0.3729	0.4031
11	0.4607	0.4684	0.4788	0.4781	0.4567	0.4777
12	0.5292	0.5315	0.5362	0.5400	0.5278	0.5350

Table 8.2 Edges, costs and BIM's values

i	edge e	$p(e)$	cost $c(e)$	BIM_i	$\alpha = BIM_i \cdot (0.9 - p(e))$	$\alpha/c(e)$
1	(1, 2)	0.6	2	0.064	0.019	0.01
2	(2, 6)	0.6	2	0.167	0.050	0.025
3	(3, 8)	0.6	2	0.167	0.050	0.025
4	(4, 11)	0.6	2	0.240	0.072	0.0.036
5	(5, 13)	0.6	2	0.242	0.073	0.036
6	(6, 7)	0.7	1	0.261	0.052	0.052
7	(7, 8)	0.7	1	0.253	0.051	0.051
8	(8, 9)	0.7	1	0.104	0.023	0.023
9	(11, 12)	0.7	1	0.218	0.044	0.044
10	(12, 13)	0.7	1	0.216	0.043	0.043

$$(6, 7), (7, 8), (4, 11), (5, 13), (11, 12), (12, 13).$$

Remark 8.1 Strictly speaking, after replacing the first edge $(6, 7)$, not all edges will have the same reliability, and that will affect the BIM's of other edges. But from a practical point of view, our heuristic approach justifies itself.

Let us now assume that **not all edges** are equally reliable and also the costs of edge reinforcement are not equal too. Namely, the reliability of edges located on a large ring is 0.7 and the cost of reinforcing an edge equals 1. For the remaining edges, the reliability is 0.6 and reinforcement cost equals 2. The initial network reliability calculated using CMC is $R = 0.479$. Our goal remains the same: to increase the reliability to 0.85, for minimal cost. Table 8.2 presents edge reliability, as well as the cost of replacing an edge by new one having $p = 0.9$. (The table shows only a part of all edges.)

The BIM's in this table are computed using the turnip algorithm (see Remark 7.4). Note that they can also be calculated by formula 5.1.4. Now the criterion for choosing an edge for reinforcement will be determined by the largest value of $BIM_e \cdot (0.9 - p(e))/c(e)$, which represents the increase in edge reliability per unit cost. Such an approach resembles the Knapsack problem algorithm. After calculations, we get the following sequence of 7 edges, resulting in network reliability $R = 0.846$.

$$(6, 7), (7, 8), (11, 12), (12, 13), (4, 11), (5, 13), (2, 6)$$

The corresponding cost is equal to $C = 10$.

Remark 8.2 Recalculating the BIMs after choosing each edge, according to the values $BIM_e \cdot (0.9 - p(e))/c(e)$, we could get a solution by a slightly lower cost.

More about various strategies of roads system predisaster design the reader can find in [2, 3].

8.3 Flow in Network with Unreliable Edges

In this section we consider flow in a network with unreliable edges. Flow networks are very important and widely used models in transportation networks and various types of supply networks, see [4, 5].

We will consider networks with independent randomly failing **directed** edges. For each edge $e = (a, b)$ directed from the node a to node b, we define the maximal flow $c(e)$ (capacity) which can be delivered from a to b along this edge. We say that the network state is UP if the maximal flow from source to sink is not less than some prescribed value Φ.

Figure 8.6 represents very simple flow network with 4 nodes and 5 edges.

It is easy to check that the maximal flow from source $s = 1$ to sink $t = 2$ equals 5. For example, it may be obtained by the following flows:

Fig. 8.6 Flow network with 4 nodes and 5 edges

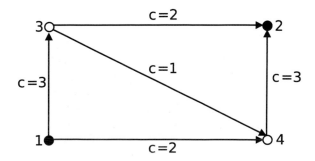

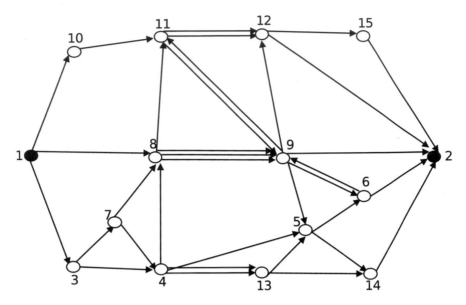

Fig. 8.7 Flow network with 15 nodes and 35 edges

$$w(1, 3) = 3, \ w(1, 4) = 2, \ w(3, 2) = 2, \ w(3, 4) = 1, \ w(4, 2) = 3.$$

Suppose that we define the UP state as the state with maximal flow no less than $\Phi = 4$. Then, if edge $(3, 2)$ is $down$, the maximal flow equals 3, and the network is $DOWN$.

Suppose now that any edge $e(x, y)$ fails with probability p and its flow capacity drops from maximal value $c(e)$ to zero. Our goal will be to estimate the probability that the maximal flow between the source node s and the sink node t will be not less than some prescribed value Φ.

Figure 8.7 presents the flow network with 35 edges and 15 nodes. 1 and 2 are the source node and the sink node, respectively.

Edge capacities are presented in Table 8.3. There are several pairs of adjacent nodes connected by two or three edges. They are denoted in the table as $(i, j - 1)$ or $(i, j) - 2$ or $(i, j) - 3$. The arrows on the edges denote the direction in which the flow is allowed to go. Network failure $(DOWN)$ is defined as the flow reduction below $\Phi = 12$. The probability of edge failure is denoted by q.

We investigate this network by means of Monte Carlo based on estimating the CD-spectrum of the network (see Chap. 6). Let us remind in short the estimation procedure. First, we consider the set of randomly ordered network element numbers— the permutations. Each simulated permutation is "destroyed" from left to right by erasing one element (edge) after another. After each destruction (edge elimination) the network state is checked by a special algorithm and the position of D-anchor is registered. After repeating this procedure M times we remember the numbers M_i of cases when the D-anchor was on the ith position and compute the cumulative

Table 8.3 Edge capacities

$e(i, j)$	$c(i, j)$	$e(i, j)$	$c(i, j)$	$e(i, j)$	$c(i, j)$
(1, 3)	8	(1, 8)	9	(1, 10)	8
(3, 4)	6	(3, 7)	6	(4, 5)	6
(4, 8)	6	(4, 13 − 1)	5	(4, 13 − 2)	2
(5, 6)	6	(5, 14)	5	(6, 2)	6
(6, 9)	3	(7, 4)	5	(7, 8)	4
—	—	(8, 4)	4	(8, 9 − 1)	5
(8, 9 − 2)	4	(8, 9 − 3)	3	(8, 11)	4
(9, 5)	6	(9, 6)	4	(9, 2)	5
(9, 11)	4	(9, 12)	5	(10, 11)	5
(11, 9)	4	(11, 12 − 1)	4	(11, 12 − 2)	2
(12, 2)	6	(12, 15)	5	(13, 14)	5
(13, 5)	5	(14, 2)	5	(15, 2)	5

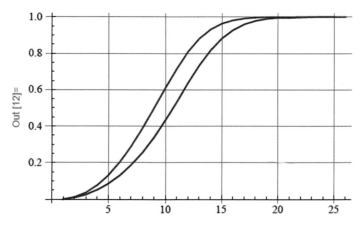

Fig. 8.8 CD-spectra for $\Phi = 12$—left curve, and $\Phi = 10$—right curve

spectrum (CD-spectrum) $y_1, \ldots, y_n$ as

$$y_1 = \frac{M_1}{M}, y_2 = \frac{M_1 + M_2}{M}, \ldots, y_n = \frac{M_1 + M_2 + \cdots + M_n}{M}$$

An important fact is that there is *no need* to check the network state on each step of the destruction process. The position of the anchor can be efficiently found by applying **bisection** algorithm which works as follows. Erase the $[n/2]$ edges of the permutation. Check the network state. If it is already *DOWN*, the anchor must be in the first $[n/2]$ positions. If not, the anchor is within remaining positions. Proceed by bisecting the part of the permutation until you locate the anchor. On the average, the location of the anchor will be found by $O(\log_2 n)$ network state checks.

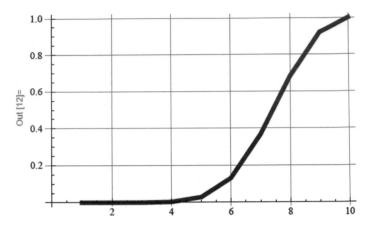

Fig. 8.9 Horizontal axis shows $x = 10p$. The vertical axis is network reliability $P(UP)$ for $\Phi = 12$

Finally, we note that the network state can be checked by applying the well-known Ford–Fulkerson classic algorithm (see [4, 5]) for calculating the max flow in a network. If this max flow is less then our limit Φ, we say that the network is *DOWN*. The CD-spectra for our network for $\Phi = 12$ and $\Phi = 10$ are presented on Fig. 8.8.

If the maximum flow is less than 10, then it is for sure less than 12, that is, the probability of *DOWN* for $\Phi = 12$ is higher. This explains why in Fig. 8.8 the left curve dominates the right one.

After knowing the CD-spectrum, network reliability is calculated by the well-known formula (6.2.3) based on the fundamental property (6.2.2) of the CD-spectrum.

The graph on Fig. 8.9 shows how depends network reliability $R = P(UP)$ on edge *up probability p.*

References

1. Barabasi, A., & Albert, R. Emergence of scaling in random networks. *Science*, 286(5439), 509–512.
2. Gertsbakh, I., & Shpungin, Y. (2011). *Network reliability and resilience*. New York: Springer.
3. Salman, S. F. (2010). Predisaster investment decisions for strengthening a highway network. *Computers and Operations Research*, *37*, 1708–1719.
4. Ford, L., & Fulkerson, D. (1962). *Flows in networks*. Santa Monica: Rand Corporation.
5. Gertsbakh, I., Rubinstein, R., Shpungin, Y., & Vaisman, R. (2014). Methods for performance analysis of stochastic flow networks. *Probability in Engineering and Information Sciences*, *28*, 21–38.

Printed in the United States
By Bookmasters